Electricity

A Brief History

By Dr Alex Bugeja, PhD

Created in part using the Qyx AI Book Creator

See "About this book" in the Introduction

Introduction

Chapter 1 The Spark of Curiosity: Early Observations of Electricity

Chapter 2 Amber's Allure: Thales and the Mystery of Static Charge

Chapter 3 Lodestones and Lightning: Separating Magnetism and Electricity

Chapter 4 The Leyden Jar: Capturing and Storing the Electric Fluid

Chapter 5 Franklin's Kite: Unveiling the Nature of Lightning

Chapter 6 Galvani vs. Volta: Animal Electricity and the First Battery

Chapter 7 The Voltaic Pile: A Continuous Flow of Electric Current

Chapter 8 Oersted's Discovery: The Intimate Link Between Electricity and Magnetism

Chapter 9 Ampère's Insights: Quantifying the Magnetic Force of Current

Chapter 10 Faraday's Breakthrough: Electromagnetic Induction

Chapter 11 The Electric Motor: From Scientific Curiosity to Practical Application

Chapter 12 The Generator: Harnessing Mechanical Energy to Produce Electricity

Chapter 13 Maxwell's Equations: Unifying Electricity, Magnetism, and Light

Chapter 14 The Dawn of Electric Lighting: Edison and the Incandescent Bulb

Chapter 15 The Battle of the Currents: AC vs. DC

Chapter 16 Tesla's Vision: Alternating Current Triumphs

Chapter 17 Hertz's Waves: Confirming Maxwell's Theory and the Birth of Radio

Chapter 18 The Electron's Discovery: Unveiling the Fundamental Unit of Charge

Chapter 19 The Rise of Electronics: Vacuum Tubes and the Amplification of Signals

Chapter 20 The Transistor Revolution: Miniaturization and the Digital Age

Chapter 21 Integrated Circuits: The Building Blocks of Modern Electronics

Chapter 22 The Power Grid: Distributing Electricity Across Nations

Chapter 23 Renewable Energy: Harnessing the Power of Nature

Chapter 24 Superconductivity: The Quest for Zero Resistance

Chapter 25 The Future of Electricity: Smart Grids, Electric Vehicles, and Beyond

Introduction

We live in an electrified world. From the moment we wake up to the moment we go to sleep, electricity powers our lives in countless ways. It lights our homes, cooks our food, entertains us, and connects us to the world. But how often do we stop to think about where electricity comes from, how it works, or the remarkable journey that led to its widespread use?

This book, "Electricity: A Brief History," aims to shed light on this fascinating story. It's a journey that spans millennia, from the earliest observations of strange phenomena to the cutting-edge technologies that define our modern era. We'll explore the sparks of curiosity that ignited the first investigations into the nature of electricity, and follow the threads of discovery as they weave through the centuries.

Our story begins in ancient times, when philosophers like Thales of Miletus pondered the mysterious properties of amber, a fossilized tree resin that could attract light objects after being rubbed. This seemingly magical effect, now known as static electricity, was a source of wonder and speculation for centuries. We'll also encounter lodestones, naturally occurring magnets whose ability to attract iron hinted at a connection between magnetism and electricity, though the true nature of this relationship wouldn't be understood until much later.

As we move forward in time, we'll witness the ingenious experiments of scientists like William Gilbert, who coined the term "electricus" and laid the groundwork for the scientific study of electricity and magnetism. We'll see how the invention of the Leyden jar in the 18th century allowed scientists to capture and store electricity for the first time, opening up new avenues of research.

The book then delves into the pivotal work of Benjamin Franklin, whose famous kite experiment demonstrated the electrical nature of lightning and revolutionized our understanding of this powerful natural phenomenon. We'll explore the debate between Luigi Galvani and Alessandro Volta over "animal electricity" and the invention of the first battery, the voltaic pile, which provided a continuous source of electric current and paved the way for a new era of experimentation.

The 19th century witnessed an explosion of discoveries and innovations related to electricity. We'll examine Hans Christian Ørsted's groundbreaking observation that an electric current could deflect a compass needle, revealing the intimate link between electricity and magnetism. We'll follow André-Marie Ampère as he quantified the magnetic force exerted by electric currents, and Michael Faraday as he discovered electromagnetic induction, the principle behind electric generators and transformers.

The book then turns to the development of practical applications of electricity, such as the electric motor and the generator. We'll see how these inventions transformed industries and societies, powering everything from factories to transportation systems. We'll also explore the work of James Clerk Maxwell, whose elegant equations unified electricity, magnetism, and light into a single framework, laying the foundation for our modern understanding of electromagnetism.

The late 19th and early 20th centuries saw the rise of electric lighting, with Thomas Edison's incandescent bulb illuminating homes and cities around the world. We'll witness the "Battle of the Currents" between Edison's direct current (DC) system and the alternating current (AC) system championed by Nikola Tesla and George Westinghouse, a struggle that ultimately shaped the infrastructure of our modern electrical grid.

The discovery of the electron by J.J. Thomson in 1897 marked a turning point in our understanding of electricity, revealing the fundamental unit of electric charge. This discovery paved the way for the development of electronics, starting with the invention of the vacuum tube, which allowed for the amplification of electrical signals and enabled the creation of technologies like radio and television.

The 20th century saw the transistor revolution, with the invention of the transistor in 1947 leading to the miniaturization of electronic devices and the dawn of the digital age. We'll explore the development of integrated circuits, which packed millions of transistors onto tiny silicon chips, enabling the creation of powerful computers and other electronic devices.

The book also examines the evolution of the power grid, the vast network of power plants, transmission lines, and distribution systems that deliver electricity to homes and businesses across nations. We'll discuss the challenges and opportunities associated with renewable energy sources, such as solar and wind power, and the quest for superconductivity, a phenomenon that could revolutionize energy transmission and storage.

Finally, we'll look to the future of electricity, exploring emerging technologies like smart grids, electric vehicles, and advanced energy storage systems. We'll consider the potential of these technologies to transform our lives and address pressing challenges like climate change.

Throughout this journey, we'll encounter a cast of brilliant scientists, inventors, and entrepreneurs who dedicated their lives to unraveling the mysteries of electricity and harnessing its power for the benefit of humanity. Their stories are a testament to the power of human curiosity, ingenuity, and perseverance.

"Electricity: A Brief History" is not just a chronicle of scientific and technological advancements. It's a story about how our understanding of the world has evolved, and how a force that was once seen as magical and mysterious has become an integral part of our daily lives. It's a story that continues to unfold, as we push the boundaries of knowledge and innovation in the quest for a more sustainable and electrified future.

About this book

The author, Dr Alex Bugeja, is the Founder & CEO of Traffikoo, a Texas company specializing in online advertising, AI tools, and

SaaS solutions. He is originally from Malta and now lives in Texas.

This book was created in part using the Qyx AI Book Creator, a project developed and maintained by Traffikoo. Qyx AI Book Creator is a powerful and affordable AI ghostwriter, capable of creating entire books on virtually any subject. It is suitable for making books to sell to others, as well as for personal use. Its books are perfectly useable as is - or as drafts for those wishing to edit them and add their own personal touches.

Besides serving as a history of electricity, we hope this book also inspires you to try out Qyx AI Book Creator for yourself.

CHAPTER ONE: The Spark of Curiosity: Early Observations of Electricity

The story of electricity begins not with wires and circuits, but with a sense of wonder. Long before humanity understood the fundamental forces governing the universe, our ancestors encountered phenomena that hinted at a hidden power, a force beyond the familiar realm of muscle and mechanics. These early encounters, though shrouded in mystery and often attributed to supernatural forces, laid the groundwork for the scientific revolution that would eventually illuminate the world, both literally and figuratively.

Imagine a world without electricity. No lights to pierce the darkness, no engines to ease our labor, no devices to connect us across vast distances. This was the reality for the vast majority of human history. People lived in rhythm with the sun, their lives dictated by the cycles of day and night. Fire was their primary source of light and heat, a precious commodity that had to be carefully tended. Yet, even in this world devoid of electrical technology, there were hints of something more.

One of the earliest recorded observations of an electrical phenomenon dates back to ancient Greece, around 600 BCE. The philosopher Thales of Miletus, often considered one of the "Seven Sages" of Greece, noticed something peculiar about amber, a fossilized tree resin prized for its beauty and used in jewelry and other decorative objects. When rubbed with a piece of cloth, amber would attract light objects like feathers or bits of straw. This seemingly magical property intrigued Thales and other thinkers of his time, though they lacked the tools and knowledge to explain it.

Thales wasn't just a curious observer; he was a pioneer of natural philosophy, a precursor to modern science. He sought to understand the world through reason and observation, rather than relying solely on myths and religious explanations. His observation of amber's attractive power, though simple, was a significant step. It was one of the first documented attempts to understand a phenomenon that we now know as static electricity.

The word "electricity" itself has its roots in this ancient fascination with amber. The Greek word for amber is "elektron," and it was from this word that the English scientist William Gilbert would later coin the term "electricus" in the 16th century. But that's getting ahead of our story.

For centuries after Thales, the strange behavior of amber remained a curiosity, a parlor trick rather than a subject of serious scientific inquiry. It was just one of many unexplained phenomena in a world that was largely a mystery. People observed lightning, a terrifying and awe-inspiring display of natural power, but they had no way of connecting it to the subtle attraction of rubbed amber.

Lightning was often attributed to the wrath of gods or other supernatural forces. The ancient Greeks believed that Zeus, the king of the gods, hurled lightning bolts from Mount Olympus. The Romans had their own god of thunder and lightning, Jupiter. In Norse mythology, Thor wielded his mighty hammer, Mjölnir, to create thunder and lightning. These myths reflected a human desire to understand and explain the powerful forces of nature, even if the explanations were based on imagination rather than empirical evidence.

Another phenomenon that hinted at a hidden force was the lodestone, a naturally magnetized piece of the mineral magnetite. Lodestones had the remarkable ability to attract iron, a property that was just as mysterious as amber's attraction to light objects. The ancient Greeks were familiar with lodestones, as were the Chinese, who may have been the first to use them for navigation in the form of a compass.

The compass, a simple device consisting of a magnetized needle that aligns itself with the Earth's magnetic field, was a revolutionary invention. It allowed sailors to navigate even when the stars were obscured by clouds, opening up new possibilities for exploration and trade. The earliest compasses were likely made by floating a lodestone on a piece of wood or cork in a bowl of water. The needle would then point north, providing a reliable way to determine direction.

The connection between the attractive power of amber and the magnetic force of the lodestone was not immediately apparent. Magnetism seemed to be a distinct phenomenon, affecting only iron and other magnetic materials. For centuries, these two forces – the electric and the magnetic – were considered separate and unrelated.

It's easy to look back from our modern perspective, with our knowledge of electrons and electromagnetic fields, and see these early observations as naive or primitive. But it's important to remember that these ancient thinkers were working with extremely limited tools and knowledge. They had no concept of atoms or subatomic particles, no instruments to measure electric charge or magnetic fields. Their observations were based on what they could see and touch, and their explanations were often based on analogy and speculation.

Despite these limitations, these early encounters with electrical and magnetic phenomena were crucial. They sparked a curiosity that would eventually lead to a deeper understanding of the natural world. They demonstrated that there were forces at play beyond the visible and the tangible, forces that could be observed, manipulated, and perhaps even harnessed.

The journey from these early observations to the complex electrical technologies of today was long and winding. It involved countless experiments, theories, and debates. It required the development of new tools and instruments, and the gradual accumulation of knowledge over centuries.

Think of it like assembling a giant jigsaw puzzle. Each observation, each experiment, each new idea was like a single piece of the puzzle. At first, the pieces didn't seem to fit together. Amber's attraction to feathers, the lodestone's pull on iron, the terrifying power of lightning – these were disparate phenomena with no apparent connection.

But slowly, gradually, as more pieces were discovered and examined, a picture began to emerge. The puzzle was far from complete, but there were hints of a larger pattern, a hidden order beneath the surface of seemingly random events.

The early Greeks, with their philosophical inquiries and their fascination with natural phenomena, laid the foundation for this long process of discovery. They may not have understood the true nature of electricity or magnetism, but they recognized that there was something there, something worth investigating.

Their legacy was not a set of answers, but a set of questions. Questions like: What is this strange force that makes amber attract light objects? Why does a lodestone attract iron? What is the nature of lightning, and is it related to these other phenomena?

These questions would echo through the centuries, driving generations of scientists and inventors to seek answers. The quest to understand electricity was not a linear path, but a complex and often messy process of trial and error, of dead ends and breakthroughs.

It's a story that highlights the importance of curiosity, observation, and the willingness to challenge existing beliefs. The early observers of electrical phenomena didn't have all the answers, but they had the courage to ask the questions. And in doing so, they set in motion a chain of events that would eventually transform the world.

The next major steps in this journey would have to wait for several centuries. The scientific revolution of the 16th and 17th centuries, with its emphasis on experimentation and mathematical analysis, would provide the necessary tools and framework for a more systematic investigation of electricity and magnetism. But the seeds of this revolution were sown in the ancient world, in the simple observations of amber and lodestone, and in the enduring human desire to understand the mysteries of the universe.

CHAPTER TWO: Amber's Allure: Thales and the Mystery of Static Charge

Thales of Miletus, a name that resonates through the corridors of ancient Greek philosophy, is often hailed as one of the "Seven Sages" of antiquity. He lived in Miletus, a bustling city on the western coast of Anatolia (present-day Turkey), during the 6th century BCE. This was a time of intellectual ferment, a period when the foundations of Western philosophy and science were being laid. Thales, a polymath, was at the forefront of this movement, his inquisitive mind probing the nature of the cosmos, the composition of matter, and the underlying principles governing the world around him.

While Thales is renowned for his contributions to geometry, astronomy, and philosophy, it is his observation of a seemingly mundane phenomenon that earns him a place in the history of electricity. This phenomenon involved amber, a fossilized tree resin that was prized in the ancient world for its beauty and used in jewelry, amulets, and other decorative objects. Amber, when rubbed with a cloth, exhibits a peculiar property: it attracts light objects like feathers, straw, or bits of dried leaves. This attraction, though subtle, was unlike anything else observed at the time. It wasn't the familiar pull of gravity, nor was it the magnetism displayed by lodestones. It was something else, something mysterious and intriguing.

Thales, with his penchant for seeking natural explanations for observed phenomena, was captivated by this peculiar behavior of amber. He wasn't content with simply attributing it to magic or the whims of the gods, as was common practice in his time. Instead, he sought a rational explanation, a principle that could account for this strange attractive force.

It's important to understand the context in which Thales was working. He didn't have access to the sophisticated instruments and experimental techniques that would later become the hallmarks of scientific inquiry. His tools were primarily his senses and his intellect. He observed the world around him, formulated hypotheses, and engaged in reasoned discourse with his peers.

In the case of amber, Thales's observations were limited to the basic phenomenon itself: the attraction of light objects after rubbing. He didn't have the means to quantify the force, measure the duration of the attraction, or investigate the nature of the "charge" involved. Nevertheless, his observation was significant because it represented a departure from the prevailing supernatural explanations of the time.

Thales's explanation for the attractive power of amber, though ultimately incorrect by modern standards, was rooted in the philosophical ideas of his time. He proposed that amber possessed a "soul" or vital force that was awakened by the friction of rubbing. This soul, he believed, was responsible for the attraction, drawing light objects towards the amber.

This concept of a soul or vital force residing within inanimate objects was not uncommon in ancient Greek thought. It was part of a broader worldview known as hylozoism, which held that all matter was, in some sense, alive or possessed a degree of consciousness. Thales himself was a proponent of hylozoism, believing that the entire universe was animated by a divine principle.

In the context of amber, the idea of a soul provided a seemingly plausible explanation for the observed attraction. The act of rubbing was thought to awaken this dormant force, allowing it to exert its influence on nearby objects. It was a simple, intuitive explanation that resonated with the prevailing philosophical ideas of the time.

It's crucial to remember that Thales's explanation was not based on empirical evidence in the modern sense. He didn't conduct controlled experiments or make precise measurements. His theory was a product of philosophical speculation, an attempt to fit the observed phenomenon into a broader understanding of the natural world.

Despite its limitations, Thales's theory represented an important step forward. It was an attempt to explain a natural phenomenon through natural principles, rather than resorting to supernatural or mystical explanations. This emphasis on naturalism was a defining characteristic of Thales's philosophical approach and a cornerstone of the emerging scientific worldview.

The word "electricity" itself owes its origin to Thales's observation of amber. The Greek word for amber is "elektron," and it was from this word that the English scientist William Gilbert would later coin the term "electricus" in the 16th century. Gilbert, who conducted extensive experiments on magnetism and static electricity, recognized the connection between the attractive properties of amber and other materials that exhibited similar behavior after being rubbed.

For centuries after Thales, the phenomenon of static electricity, as we now call it, remained largely a curiosity. It was a parlor trick, a source of amusement rather than a subject of serious scientific investigation. People observed that other materials, such as glass and certain types of gemstones, could also attract light objects after being rubbed, but the underlying mechanism remained a mystery.

The lack of progress in understanding static electricity during this period can be attributed to several factors. First, there was no theoretical framework within which to understand the phenomenon. The concept of electric charge, let alone the existence of subatomic particles like electrons, was far beyond the grasp of ancient and medieval thinkers.

Second, there were no instruments or techniques available to study static electricity in a systematic way. The ability to generate, store, and measure electric charge would have to wait for the development of more sophisticated technologies in later centuries.

Third, static electricity, unlike magnetism, was a relatively weak and transient phenomenon. The attractive force generated by rubbing amber was short-lived and only affected very light objects. This made it difficult to study and less compelling than the more powerful and persistent force of magnetism exhibited by lodestones.

Despite these limitations, the observation of amber's attractive power remained a persistent thread in the tapestry of scientific inquiry. It was a reminder that there were forces at play in the universe beyond those that were immediately apparent. It was a puzzle that would eventually be solved, but not until the tools and knowledge were available to tackle it.

The story of Thales and amber is not just a historical footnote. It's a testament to the power of observation, the importance of seeking natural explanations, and the enduring human curiosity about the world around us. Thales may not have understood the true nature of static electricity, but he recognized that there was something there, something worth investigating.

His observation, though simple, was a spark that ignited a long and winding journey of discovery. It was a journey that would eventually lead to a profound understanding of electricity and its fundamental role in the universe. It's a journey that continues to this day, as we probe the mysteries of the atom and harness the power of electricity to shape our world in ever more remarkable ways.

Thales' legacy is not confined to the realm of electricity. He is considered one of the founders of Western philosophy and a pioneer of the scientific method. His emphasis on reason, observation, and natural explanations laid the groundwork for the intellectual revolution that would transform the world.

He was a key figure in the transition from a mythical worldview to a more rational and empirical one. He challenged the prevailing explanations of natural phenomena based on the whims of the gods and sought instead to understand the world through natural principles.

Thales's contributions to geometry and astronomy were equally groundbreaking. He is credited with introducing the study of geometry to Greece, and he is said to have predicted a solar eclipse in 585 BCE, a feat that would have required a sophisticated understanding of celestial mechanics.

His philosophical ideas, particularly his theory that water was the fundamental substance of all things, were influential in shaping the development of early Greek thought. While his specific theories about the composition of matter have been superseded by modern science, his emphasis on identifying underlying principles and unifying concepts remains a cornerstone of scientific inquiry.

The story of Thales and amber serves as a reminder that scientific progress is often a slow and iterative process. It's a journey that begins with simple observations, sparks of curiosity that ignite a desire to understand the world around us. It's a journey that requires patience, persistence, and a willingness to challenge existing beliefs.

Thales, with his inquisitive mind and his commitment to natural explanations, set in motion a chain of events that would eventually lead to a profound transformation in our understanding of the universe. His story is a testament to the power of human curiosity and the enduring quest for knowledge that has driven scientific progress throughout history. It is a legacy that continues to inspire us today, as we continue to explore the mysteries of the cosmos and harness the power of electricity to shape a brighter future.

CHAPTER THREE: Lodestones and Lightning: Separating Magnetism and Electricity

For centuries, the subtle attraction of rubbed amber and the powerful pull of the lodestone were two sides of the same mysterious coin. Both phenomena hinted at hidden forces beyond the realm of everyday experience, forces that seemed to defy explanation. While amber, with its gentle allure for light objects, became associated with what we now know as static electricity, the lodestone, a naturally magnetized piece of the mineral magnetite, held sway as the enigmatic embodiment of magnetism. These two forces, though seemingly distinct in their manifestations, were often conflated, their true relationship obscured by a lack of understanding and the absence of tools to probe their underlying nature.

The lodestone, like amber, had been known since antiquity. Its name derives from the Old English word "lode," meaning "way" or "journey," a testament to its crucial role in the development of the compass and its impact on navigation. Imagine the sense of wonder and perhaps even fear that these naturally occurring magnets must have inspired in early civilizations. Rocks that could mysteriously attract iron, pulling metal objects towards them with an invisible force, seemed to possess a magical or even supernatural power.

The ancient Greeks, always keen to unravel the mysteries of the natural world, were familiar with lodestones. They found deposits of this magnetic mineral near the city of Magnesia in Thessaly, from which the term "magnetism" itself is derived. Like amber, the lodestone became a subject of philosophical speculation. Thales of Miletus, the same philosopher who pondered the attractive power of amber, also contemplated the nature of magnetism. He, like many of his time, believed that the lodestone possessed a "soul" or animating force that was responsible for its ability to attract iron.

The Chinese were also aware of the lodestone's unique properties. Evidence suggests that they may have been the first to harness its power for navigation, possibly as early as the 4th century BCE. Early Chinese compasses, known as "south-pointing needles," were often fashioned from lodestones carved into the shape of spoons or ladles. These spoons, when placed on a smooth surface or floated on water, would align themselves with the Earth's magnetic field, with the handle pointing south.

The invention of the compass was a pivotal moment in human history. It revolutionized navigation, enabling sailors to venture far beyond the sight of land and opening up new routes for trade and exploration. The ability to determine direction reliably, even under cloudy skies or in unfamiliar waters, was a game-changer. It facilitated the great voyages of discovery that would eventually connect the world in unprecedented ways.

Despite the practical application of the compass, the underlying nature of magnetism remained a mystery for centuries. The force exerted by a lodestone seemed fundamentally different from the attraction observed in rubbed amber. Magnetism was a more powerful and persistent force, capable of attracting iron objects from a distance and without the need for rubbing. Furthermore, magnetism appeared to be specific to certain materials, primarily iron and its alloys, while the attractive power of amber seemed to be more universal, albeit weaker.

These differences led some thinkers to speculate that magnetism and the amber effect were distinct phenomena. However, in the absence of a clear understanding of either force, the two were often grouped together as examples of occult or hidden powers. They were both seen as manifestations of a mysterious force that resided within certain materials, a force that could be awakened or manipulated under specific conditions.

The scientific study of magnetism and electricity began to diverge more definitively during the Renaissance, a period of renewed interest in classical learning and a burgeoning spirit of scientific inquiry. One of the key figures in this transition was William

Gilbert, an English physician and natural philosopher who lived in the late 16th and early 17th centuries. Gilbert is often considered the "father of magnetism" for his groundbreaking work on the subject.

Gilbert's magnum opus, "De Magnete, Magneticisque Corporibus, et de Magno Magnete Tellure" (On the Magnet and Magnetic Bodies, and on the Great Magnet the Earth), published in 1600, was the first comprehensive scientific treatise on magnetism. In this work, Gilbert meticulously documented his experiments and observations, laying the foundation for the scientific study of magnetism and electricity.

Gilbert conducted extensive experiments with lodestones and other magnetic materials. He carefully studied the properties of magnets, including their poles, their ability to attract and repel each other, and their influence on iron. He demonstrated that the Earth itself behaves like a giant magnet, with its own magnetic poles and magnetic field. This insight explained the behavior of the compass needle, which aligns itself with the Earth's magnetic field, always pointing towards the magnetic north.

One of Gilbert's most significant contributions was his clear distinction between magnetism and the attractive force exhibited by rubbed amber. He recognized that these were two fundamentally different phenomena, governed by different principles. He coined the term "electricus," derived from the Greek word "elektron" (amber), to describe the attractive force produced by rubbing materials like amber, glass, and resin.

Gilbert's experiments showed that while magnets attracted only iron and a few other materials, "electrics" could attract a wide range of light objects. He also observed that the electric attraction was temporary and easily dissipated, while the magnetic force was more persistent. Furthermore, he noted that heat destroyed the magnetism of a lodestone, but it enhanced the electric attraction of amber.

Through these careful observations and experiments, Gilbert established that magnetism and electricity were distinct forces. He laid the groundwork for the separate study of these two phenomena, paving the way for future researchers to delve deeper into their respective properties and underlying mechanisms.

Gilbert's work was a major turning point in the history of science. It marked a shift from the speculative and often mystical approach of earlier thinkers to a more empirical and systematic approach based on observation, experimentation, and careful analysis. His meticulous documentation of his findings and his clear articulation of the differences between magnetism and electricity set a new standard for scientific inquiry.

However, despite Gilbert's insights, the true nature of both electricity and magnetism remained elusive. He believed that both forces were due to "effluvia," invisible fluids or emanations that were released by materials under certain conditions. While this theory was ultimately incorrect, it represented a significant step forward from the earlier animistic explanations that attributed these forces to souls or vital forces.

While Gilbert was making strides in distinguishing magnetism from electricity, another natural phenomenon was captivating the minds of those seeking to understand hidden powers: lightning. Lightning, with its awesome power and destructive potential, had always been a source of fear and fascination. In many cultures, it was attributed to the wrath of gods or other supernatural beings. The ancient Greeks believed that Zeus hurled lightning bolts from Mount Olympus, while the Norse god Thor wielded his hammer Mjölnir to create thunder and lightning.

These myths reflected a human desire to explain the terrifying and unpredictable nature of lightning. It was a force that could strike without warning, setting fires, destroying property, and even taking lives. It was clearly a force to be reckoned with, and its connection to the heavens seemed to suggest a divine or supernatural origin.

The idea that lightning might be related to the subtle electrical effects observed in rubbed amber was not immediately obvious. Lightning was a large-scale, dramatic phenomenon, while the attraction of amber was a small-scale, subtle effect. However, as scientists began to study electricity more systematically, they started to notice some intriguing parallels between the two.

One of the first to suggest a possible connection between lightning and electricity was William Wall, an English physician who wrote about his observations in a paper published in 1708. Wall noticed that when he rubbed a piece of amber vigorously, it produced not only the familiar attractive force but also small sparks and crackling sounds. These sparks, he observed, bore a striking resemblance to miniature lightning flashes.

Wall's observation was a remarkable insight, a leap of imagination that connected two seemingly disparate phenomena. He speculated that lightning might be nothing more than a large-scale version of the sparks produced by rubbed amber. This was a bold hypothesis, one that challenged the prevailing view of lightning as a supernatural phenomenon.

Wall's ideas were initially met with skepticism. The notion that the awesome power of lightning could be related to the tiny sparks produced in a laboratory seemed far-fetched to many. However, his work planted a seed of doubt, a suggestion that perhaps lightning was not so different from other electrical phenomena after all.

The quest to understand the nature of lightning and its relationship to electricity would continue throughout the 18th century, culminating in the groundbreaking experiments of Benjamin Franklin, which will be covered in detail in later chapters. But the seeds of this investigation were sown in the careful observations of Gilbert and Wall, who dared to question the prevailing wisdom and to seek natural explanations for the mysteries of the universe.

The separation of magnetism and electricity was a crucial step in the development of our understanding of both forces. It allowed scientists to focus on the specific properties of each phenomenon, to develop tools and techniques for studying them, and to formulate theories about their underlying nature. This process of differentiation, of recognizing that seemingly similar phenomena are in fact distinct, is a fundamental aspect of scientific progress. It's like sorting through a jumbled collection of puzzle pieces and realizing that they belong to different puzzles, each with its own unique picture to reveal.

CHAPTER FOUR: The Leyden Jar: Capturing and Storing the Electric Fluid

The early 18th century was a time of burgeoning interest in the curious phenomenon of electricity. The foundations laid by William Gilbert, with his careful distinction between magnetism and the "electric" force, had opened up new avenues of inquiry. Scientists, or "natural philosophers" as they were often called, were eager to explore this mysterious force, to understand its properties, and to find ways to generate and manipulate it. However, they faced a significant challenge: electricity, as generated by rubbing materials like amber or glass, was fleeting and difficult to study. It dissipated quickly, and there was no way to accumulate or store it for later use.

This all changed in the mid-1740s with the invention of a remarkable device that would revolutionize the study of electricity: the Leyden jar. This simple yet ingenious invention, developed independently by two researchers working on opposite sides of the North Sea, provided a means to capture and store a substantial amount of electric charge, paving the way for more controlled and sustained experiments.

The first iteration of the Leyden jar was created by Ewald Georg von Kleist, a German cleric and amateur scientist, in the Pomeranian town of Cammin (now Kamień Pomorski in Poland). In October 1745, von Kleist was experimenting with an electrostatic generator, a device that produced static electricity through friction. These early generators typically consisted of a rotating glass sphere or cylinder that was rubbed by hand or with a piece of leather, creating a buildup of electric charge on the surface of the glass.

Von Kleist's setup involved a glass globe, a rotating crank, and a gun barrel suspended by silk threads that served as a conductor, drawing the electric charge from the glass. His aim was to see if he could collect the "electric virtue" or fluid in a vial. He inserted a nail into a small medicine bottle filled with alcohol and connected the nail to the gun barrel with a wire. He held the bottle in one hand and charged it using his electrostatic generator.

After a while, he removed the nail from the gun barrel, still holding the vial in one hand, and touched the nail with his other hand. To his astonishment, he received a powerful shock, far more intense than anything he had experienced before. He described the experience as a jolt that threw his arm and shoulder out of joint.

Von Kleist had inadvertently created the first capacitor, a device capable of storing electrical energy. The glass of the medicine bottle acted as an insulator, separating the alcohol (a conductor) inside from his hand (another conductor) on the outside. The electrostatic generator had transferred charge to the nail and the alcohol, while his hand, in contact with the outside of the bottle, had accumulated an opposite charge. When he touched the nail, he created a path for the stored charge to flow, resulting in the powerful shock.

Von Kleist, however, did not fully grasp the significance of his discovery. He initially believed that the alcohol was essential to the effect, and he didn't recognize the importance of the two conductors separated by an insulator. He shared his findings with other scientists, but his descriptions were somewhat unclear, and others had difficulty replicating his results.

Meanwhile, across the North Sea in the Dutch city of Leiden, Pieter van Musschenbroek, a professor of physics at the University of Leiden, was independently working along similar lines. Musschenbroek was a prominent scientist of his time, known for his work in physics and mathematics. He was also deeply interested in the study of electricity and had been conducting experiments with electrostatic generators.

In early 1746, Musschenbroek, assisted by his student Andreas Cunaeus, made a similar discovery. They were attempting to collect electric charge in a glass jar filled with water. Cunaeus held the jar in his hand while Musschenbroek charged it using an electrostatic generator. A wire connected to the generator was suspended in the water inside the jar.

After charging the jar, Cunaeus, not realizing that he himself was part of the circuit, touched the wire with his other hand. He received a tremendous shock, much stronger than any he had previously encountered. Musschenbroek, intrigued by this powerful effect, repeated the experiment himself and experienced a similar jolt. He famously wrote to a colleague, describing the shock as so intense that he "would not take a second shock for the kingdom of France."

Unlike von Kleist, Musschenbroek and his colleagues quickly recognized the importance of their discovery. They realized that the jar was acting as a storage device for electricity, accumulating a large amount of charge that could be discharged suddenly and powerfully. They also understood the crucial role of the two conductors (the water and the hand) separated by the insulating glass.

News of Musschenbroek's "Leiden jar," as it became known, spread rapidly throughout Europe. It created a sensation in scientific circles and beyond. The ability to store and discharge electricity at will opened up a whole new world of experimentation. Scientists could now study the properties of electricity in a more controlled and systematic way. They could investigate the effects of electric shocks on various materials, explore the nature of electric charge, and even delve into the physiological effects of electricity on living organisms.

The Leyden jar became an essential tool in every electrical researcher's laboratory. It was used in countless experiments, leading to a deeper understanding of the nature of electricity. It also became a popular attraction at public demonstrations and lectures, where audiences marveled at the sparks and shocks produced by this remarkable device.

The basic design of the Leyden jar evolved over time. The water inside the jar was eventually replaced with metal foil, which provided a more efficient conductor. Similarly, the outside of the jar was coated with another layer of foil, replacing the hand as the second conductor. This improved design, with two conductive layers separated by the insulating glass, became the standard form of the Leyden jar.

The Leyden jar was not just a scientific instrument; it was also a source of entertainment and wonder. Public demonstrations of its power became a popular form of spectacle. Showmen would charge up Leyden jars and deliver shocks to volunteers, often with dramatic and humorous results. These demonstrations, while sometimes frivolous, helped to popularize the study of electricity and to generate public interest in this new and exciting field.

One of the most famous demonstrations involving Leyden jars was conducted by Jean-Antoine Nollet, a French clergyman and physicist, before King Louis XV at the Palace of Versailles. Nollet assembled a chain of 200 monks, each holding an iron wire connected to the next person in line. He then connected one end of the chain to a charged Leyden jar and the other end to ground. When the circuit was completed, all 200 monks simultaneously jumped into the air, convulsed by the electric shock. This dramatic demonstration, while perhaps a bit cruel, certainly captured the attention of the royal court and further fueled the public's fascination with electricity.

The Leyden jar also played a role in early investigations into the nature of lightning. Scientists began to suspect that lightning was simply a large-scale electrical discharge, similar to the sparks produced by electrostatic generators and Leyden jars. This idea was further supported by the observation that lightning could electrify objects, just like the sparks from a Leyden jar.

The Leyden jar was a crucial stepping stone in the journey towards understanding and harnessing the power of electricity. It provided a means to store and manipulate electric charge, allowing for more controlled experiments and a deeper understanding of its properties. It was a simple device, yet it had a profound impact on the development of electrical science.

The invention of the Leyden jar was a pivotal moment in the history of electricity. It marked a transition from the study of fleeting, static charges to the investigation of stored and controllable electrical energy. This breakthrough paved the way for the development of more sophisticated electrical devices and laid the groundwork for the electrical revolution that would transform the world in the centuries to come. Without the ability to capture and store electricity, the remarkable progress that followed would have been impossible. The Leyden jar, though a simple device by today's standards, was a giant leap forward in humanity's quest to understand and harness the power of electricity.

CHAPTER FIVE: Franklin's Kite: Unveiling the Nature of Lightning

The Leyden jar, with its ability to store and release substantial amounts of electric charge, had ignited a firestorm of scientific curiosity across Europe. Yet, across the Atlantic, in the burgeoning city of Philadelphia, a polymath named Benjamin Franklin was about to take the study of electricity to a whole new level. Franklin, a man of diverse talents – a printer, writer, inventor, diplomat, and statesman – was also a keen observer of the natural world. His insatiable curiosity and his penchant for practical experimentation would lead him to conduct one of the most famous, and arguably most dangerous, experiments in the history of science: the kite experiment.

By the mid-18th century, the suspicion that lightning was a form of electricity had grown stronger. The sparks produced by electrostatic generators and Leyden jars bore an uncanny resemblance to miniature lightning flashes. Both phenomena produced light, crackling sounds, and even a similar smell (now known to be ozone). Furthermore, both could electrify objects, and both could deliver a shock, albeit on vastly different scales.

Several scientists, including William Wall, whose earlier observations of sparks from rubbed amber had hinted at a connection to lightning, and Jean-Antoine Nollet, who had famously shocked 200 monks with a Leyden jar, had speculated about the electrical nature of lightning. However, no one had yet devised a way to prove this hypothesis definitively. This is where Benjamin Franklin entered the scene.

Franklin, though geographically distant from the centers of European scientific activity, was well-versed in the latest research on electricity. He had obtained a Leyden jar and other electrical apparatus and had been conducting his own experiments, often with the help of a group of like-minded individuals who formed what they called the "Junto," a discussion group dedicated to intellectual and social improvement.

Franklin's experiments with the Leyden jar and other devices led him to develop a new theory of electricity. He proposed that electricity was not created by friction, as was commonly believed, but rather that it was a single "fluid" that existed in all matter. He hypothesized that objects could have either an excess or a deficiency of this fluid, which he termed "positive" and "negative" charges, respectively. This "one-fluid" theory, though not entirely accurate by modern standards, was a significant departure from the prevailing "two-fluid" theory, which posited the existence of two distinct types of electrical fluid.

Franklin's theory also introduced the concept of conservation of charge, the idea that the total amount of electrical charge in a closed system remains constant. He believed that when two objects were rubbed together, one object gained an excess of the electrical fluid (becoming positively charged), while the other lost an equal amount (becoming negatively charged). This concept of charge conservation remains a fundamental principle of electricity to this day.

Franklin's theoretical work provided a framework for understanding a wide range of electrical phenomena. He explained how the Leyden jar worked, how objects could be charged by induction, and how pointed objects tended to attract or discharge electricity more readily than blunt ones. This last observation would prove crucial in his later work on lightning.

In 1750, Franklin published a paper titled "Opinions and Conjectures concerning the Properties and Effects of the Electrical Matter, arising from Experiments and Observations made at Philadelphia." In this paper, he outlined his one-fluid theory of electricity and proposed an experiment to test the hypothesis that lightning was an electrical discharge.

Franklin's proposed experiment was both ingenious and audacious. He suggested that a person standing on an insulating platform, inside a sentry box during a thunderstorm, could draw "electric fire" from a cloud by using a long, pointed iron rod. The rod, he believed, would attract the electrical charge from the cloud, and the person could then draw sparks from the rod, demonstrating the electrical nature of lightning.

This experiment, known as the "sentry box experiment," was fraught with danger. The person conducting the experiment would be exposed to the full force of the thunderstorm, risking electrocution if the rod were struck by lightning. Despite the risks, several scientists in Europe attempted to carry out Franklin's experiment.

The first successful attempt was made in May 1752 by Thomas-François Dalibard, a French scientist, at Marly-la-Ville, near Paris. Dalibard, following Franklin's instructions, erected a 40-foot-tall iron rod and, during a thunderstorm, managed to draw sparks from it, confirming Franklin's hypothesis. News of Dalibard's success spread quickly, and other scientists soon replicated the experiment, further solidifying the connection between lightning and electricity.

Tragically, not all attempts were successful. In July 1753, Georg Wilhelm Richmann, a Swedish physicist working in St. Petersburg, Russia, was killed when he was struck by lightning while performing the sentry box experiment. Richmann's death served as a stark reminder of the dangers involved in these early electrical investigations.

Meanwhile, in Philadelphia, Franklin, unaware of the experiments being conducted in Europe, decided to carry out his own version of the experiment. However, instead of using a sentry box and a long iron rod, he devised a more practical, and perhaps slightly safer, method: a kite.

Franklin's kite experiment, as it has come to be known, is shrouded in some uncertainty. He never published a detailed account of the experiment himself, and the only contemporary description comes from a brief passage in a letter he wrote to his friend Peter

Collinson in October 1752. This lack of detail has led to some debate about the exact nature of the experiment and whether it even took place as commonly described.

According to the popular account, Franklin, assisted by his son William, flew a silk kite during a thunderstorm. The kite was fitted with a pointed wire at the top to attract electrical charge from the clouds. A hemp string, which became conductive when wet, connected the kite to a key near the bottom. An insulating silk ribbon was attached to the hemp string, and Franklin held this ribbon to avoid being directly connected to the potentially dangerous electrical charge.

As the thunderstorm approached, Franklin observed that the loose fibers of the hemp string stood erect, indicating that they were charged. He then brought his knuckle near the key and drew a spark, demonstrating that the kite was indeed collecting electrical charge from the atmosphere. This spark, similar to those produced by a Leyden jar, provided strong evidence that lightning was an electrical phenomenon.

It's important to note that Franklin did not get struck by lightning during the experiment, as is sometimes depicted in popular illustrations. He was careful to insulate himself from the kite and to avoid direct contact with the charged string. The experiment was still dangerous, but Franklin's precautions likely prevented him from suffering the same fate as Richmann.

The kite experiment, though perhaps not as dramatic as often portrayed, was a landmark achievement. It provided compelling evidence for the electrical nature of lightning and further validated Franklin's one-fluid theory of electricity. It also captured the public imagination, cementing Franklin's reputation as a leading scientific figure of his time.

Following the kite experiment, Franklin turned his attention to the practical application of his discovery. He reasoned that if lightning was indeed electricity, then it should be possible to protect buildings from its destructive power by using pointed rods to attract and safely discharge the electrical energy. This line of thinking led him to invent the lightning rod.

The lightning rod, a simple yet ingenious device, consisted of a pointed metal rod mounted on the highest point of a building and connected to the ground by a conductive wire. Franklin's theory was that the pointed rod would attract the electrical charge from a thundercloud, gradually and silently discharging it to the ground, thus preventing a sudden and destructive lightning strike.

Franklin's invention was quickly adopted throughout the American colonies and Europe. It proved to be highly effective in protecting buildings from lightning damage, saving countless lives and properties. The lightning rod was a testament to Franklin's ability to translate scientific discoveries into practical applications that benefited society.

The impact of Franklin's work on electricity extended far beyond the kite experiment and the lightning rod. His one-fluid theory, though later modified, provided a conceptual framework for understanding electrical phenomena. His concept of conservation of charge remains a fundamental principle of electromagnetism. His experiments and observations inspired a generation of scientists to delve deeper into the mysteries of electricity, paving the way for the remarkable advancements of the 19th and 20th centuries.

Franklin's contributions to the field of electricity were recognized both during his lifetime and in the centuries that followed. He was elected a Fellow of the Royal Society of London, a prestigious honor bestowed upon the leading scientists of the day. His work was translated into numerous languages and widely disseminated throughout Europe and the Americas. He became an international celebrity, admired not only for his scientific achievements but also for his wit, wisdom, and political leadership.

Benjamin Franklin's legacy extends far beyond his scientific endeavors. He was a key figure in the American Revolution, a signer of the Declaration of Independence, and a delegate to the

Constitutional Convention. He played a crucial role in securing French support for the American cause, and he served as the first United States Ambassador to France. He was a prolific writer, known for his "Poor Richard's Almanack," his autobiography, and numerous essays on a wide range of topics.

Yet, despite his many accomplishments in other fields, Franklin's work on electricity remains one of his most enduring legacies. His kite experiment, though perhaps more legend than reality, has become an iconic symbol of scientific curiosity and ingenuity. His invention of the lightning rod, a practical application of his scientific discoveries, stands as a testament to his commitment to using knowledge for the betterment of humanity. Franklin's contributions to the understanding of electricity laid the groundwork for the electrical revolution that would transform the world in ways he could scarcely have imagined. His story serves as an inspiration to scientists, inventors, and anyone who seeks to unravel the mysteries of the universe and to harness the power of nature for the benefit of humankind.

CHAPTER SIX: Galvani vs. Volta: Animal Electricity and the First Battery

The late 18th century was a period of intense fascination with electricity. Benjamin Franklin's groundbreaking work had shown that lightning was a form of electricity and had introduced the concept of positive and negative charges. The Leyden jar allowed scientists to store and study electricity in a controlled manner. Yet, despite these advancements, the fundamental nature of electricity remained a mystery. Was it a fluid, as Franklin had proposed? Was it a single entity or two opposing forces? And what was its relationship to living organisms?

These questions captivated the minds of scientists across Europe, and one of the most intriguing lines of inquiry emerged from the work of Luigi Galvani, an Italian physician and physicist. Galvani's experiments with frog legs would spark a scientific controversy that would not only deepen our understanding of electricity but also lead to the invention of the first battery, a device that would revolutionize the study of electricity and pave the way for countless technological advancements.

Galvani, a professor of anatomy at the University of Bologna, had long been interested in the phenomenon of "animal electricity," the idea that electricity played a role in the functioning of living organisms. This concept was not entirely new. For centuries, people had observed the electric shocks produced by certain fish, such as the electric eel and the torpedo ray. These creatures seemed to possess an innate ability to generate electricity, a power that was both fascinating and potentially dangerous.

In the 1770s, Galvani began a series of experiments to investigate the effects of electricity on animal tissues. He was particularly interested in the phenomenon of muscular contraction and the role that nerves played in transmitting signals from the brain to the muscles. He used a variety of techniques to stimulate the muscles of dissected animals, including applying sparks from electrostatic generators and Leyden jars.

Galvani's experiments took a dramatic turn in the early 1780s when he made a serendipitous observation. He was dissecting a frog on a table where he had also been conducting experiments with an electrostatic generator. One of his assistants touched the exposed sciatic nerve of the frog with a metal scalpel, and, to their astonishment, the frog's legs twitched violently. This happened when the electrostatic machine produced a spark at the same time.

Galvani was intrigued by this unexpected result. He repeated the experiment numerous times, varying the conditions and using different materials. He found that he could produce the same muscular contractions by touching the frog's nerve with a metal scalpel while a spark was generated nearby, even if the scalpel was not directly connected to the electrostatic machine. He also observed that the effect was stronger when he used two different metals, such as iron and brass, connected together to form a metallic arc.

Galvani's observations led him to believe that he had discovered a new form of electricity, one that was inherent to living organisms. He called this "animal electricity" and proposed that it was a vital force, a fluid that flowed through the nerves and was responsible for muscle movement and other life processes. He hypothesized that the frog's legs acted as a kind of natural Leyden jar, storing this animal electricity, and that the metallic arc served to discharge the stored electricity, causing the muscles to contract.

In 1791, Galvani published his findings in a landmark work titled "De Viribus Electricitatis in Motu Musculari Commentarius" (Commentary on the Effect of Electricity on Muscular Motion). In this treatise, he described his experiments in detail and presented his theory of animal electricity. He argued that the animal electricity was distinct from the "artificial" electricity produced by friction in electrostatic generators or stored in Leyden jars.

Galvani's work created a sensation in the scientific community. His experiments were replicated by scientists across Europe, and his theory of animal electricity sparked a lively debate. Some embraced his ideas, seeing them as a potential key to understanding the nature of life itself. Others were more skeptical, questioning his interpretation of the results and seeking alternative explanations.

One of the most prominent critics of Galvani's theory was Alessandro Volta, a professor of physics at the University of Pavia, just south of Milan in Italy. Volta was a leading expert on electricity and had made significant contributions to the field, including the invention of the electrophorus, a device for generating static electricity. He was initially intrigued by Galvani's experiments and even replicated them himself. However, he soon began to doubt Galvani's interpretation of the results.

Volta's skepticism stemmed from his belief that the electricity involved in Galvani's experiments was not a unique "animal electricity" but rather the same kind of electricity that was produced by friction or stored in Leyden jars. He suspected that the frog's legs were not generating electricity but merely acting as sensitive detectors of it. He argued that the crucial factor in Galvani's experiments was not the animal tissue itself but the contact between the two different metals.

Volta began a series of experiments to test his hypothesis. He replaced the frog's legs with other materials, such as moist paper or cardboard, and found that he could still produce electrical effects by simply placing them between two different metals. He also discovered that the strength of the electrical effect depended on the types of metals used and the nature of the moist material between them.

Through these experiments, Volta came to a groundbreaking conclusion: the electricity in Galvani's experiments was not coming from the animal tissue but from the contact between the two different metals. He proposed that the contact between dissimilar metals generated a small electrical current, and that the frog's legs were simply acting as a very sensitive electroscope, detecting this current and contracting in response.

Volta's theory, known as the "contact theory" of electricity, was a radical departure from Galvani's concept of animal electricity. It suggested that electricity was not a unique property of living organisms but a fundamental physical phenomenon that could be generated by inorganic materials. This was a major shift in thinking, one that would have profound implications for the future of electrical science.

The debate between Galvani and Volta, often referred to as the "frog's legs controversy," became one of the most famous scientific disputes of the 18th century. It was not just a disagreement about the interpretation of experimental results; it was a clash between two fundamentally different worldviews. Galvani, the physician and anatomist, saw electricity as intimately connected to the nature of life, a vital force that animated living organisms. Volta, the physicist, viewed electricity as a physical phenomenon, governed by the same laws that applied to inanimate matter.

To further test his contact theory, Volta devised a new and ingenious experiment. He constructed a stack of alternating discs of two different metals, typically zinc and silver, separated by pieces of cardboard or cloth soaked in a salt solution. This stack, which became known as the "voltaic pile," was the first true electric battery.

The voltaic pile was a revolutionary invention. When Volta connected wires to the top and bottom of the pile, he found that he could produce a continuous flow of electric current. This was unlike anything that had been achieved before. Electrostatic generators and Leyden jars could produce only brief discharges of static electricity. The voltaic pile, on the other hand, provided a steady and sustained source of current.

Volta's invention was a direct consequence of his contact theory. He believed that the electrical current was generated by the contact between the dissimilar metals, and that the salt solution served as a conductor, facilitating the flow of electricity. The more pairs of discs he added to the pile, the stronger the current became.

The voltaic pile was a watershed moment in the history of electricity. It provided scientists with a powerful new tool for studying the properties of electric current and its effects on various materials. It opened up a whole new realm of experimentation, leading to a deeper understanding of the relationship between electricity and chemical reactions.

Volta announced his invention in 1800 in a letter to Sir Joseph Banks, the president of the Royal Society of London. The letter, titled "On the Electricity excited by the mere Contact of conducting Substances of different Kinds," described the construction of the voltaic pile and its ability to produce a continuous electric current.

News of Volta's invention spread rapidly throughout Europe, creating a wave of excitement in the scientific community. The voltaic pile was quickly replicated by scientists across the continent, and its implications were immediately recognized. It provided a means to generate electricity at will, without the need for cumbersome electrostatic machines or Leyden jars. It allowed for sustained experiments on the properties of electric current, paving the way for a new era of electrical research.

The invention of the battery also had a profound impact on the debate between Galvani and Volta. While Galvani's supporters continued to defend the concept of animal electricity, the voltaic pile provided strong evidence for Volta's contact theory. It demonstrated that electricity could be generated by purely inorganic means, without any involvement of animal tissues.

Despite the success of the voltaic pile, the controversy surrounding animal electricity did not disappear overnight. Galvani himself never fully accepted Volta's theory, and his nephew, Giovanni Aldini, continued to conduct experiments in an attempt to vindicate his uncle's ideas. Aldini's experiments, some of which involved applying electrical currents to the bodies of recently executed criminals, were both macabre and controversial. While they did demonstrate that electricity could stimulate muscle contractions, they did not prove the existence of a unique animal electricity.

The debate between Galvani and Volta eventually faded as the scientific community came to accept Volta's contact theory as the correct explanation for the generation of electricity in the voltaic pile. However, Galvani's work was not entirely forgotten. His pioneering studies of the effects of electricity on animal tissues laid the groundwork for the field of electrophysiology, the study of the electrical activity of cells and tissues.

Galvani's legacy extends beyond his specific theories about animal electricity. He is remembered for his meticulous experimental approach, his careful observations, and his willingness to challenge existing ideas. His work inspired a generation of scientists to investigate the role of electricity in living organisms, leading to important discoveries about the nervous system, muscle function, and the electrical nature of the heart.

Volta's invention of the battery, on the other hand, had a more immediate and far-reaching impact. It provided a crucial tool for the development of electrical science and technology. The ability to generate a continuous flow of electric current opened up new possibilities for experimentation and innovation. It laid the groundwork for the development of electromagnetism, electric lighting, and countless other technologies that would transform the world in the 19th and 20th centuries.

The story of Galvani and Volta is a fascinating example of how scientific progress often arises from disagreement and debate. Their contrasting views on the nature of electricity, fueled by careful experimentation and reasoned argument, led to a deeper understanding of this fundamental force. Galvani's focus on the biological aspects of electricity and Volta's emphasis on the physical principles involved ultimately converged to create a more complete picture of the electrical nature of the universe. Their work reminds us that science is not a solitary pursuit but a collaborative endeavor, where different perspectives and approaches can lead to profound insights and transformative discoveries.

The legacy of Galvani and Volta is not merely a historical curiosity. Their work continues to resonate today in fields ranging from neuroscience to materials science. The principles behind the voltaic pile are still used in modern batteries, albeit in more sophisticated forms. The study of bioelectricity, which Galvani pioneered, remains a vibrant and important field of research. Their story serves as a reminder of the power of scientific inquiry, the importance of challenging established ideas, and the enduring human quest to understand the fundamental forces that govern our world.

CHAPTER SEVEN: The Voltaic Pile: A Continuous Flow of Electric Current

Alessandro Volta's invention of the voltaic pile in 1800 was a watershed moment in the history of electricity. It was the first device capable of producing a continuous flow of electric current, a breakthrough that would revolutionize the study of electricity and pave the way for countless technological advancements. Before the voltaic pile, scientists were limited to studying the fleeting effects of static electricity generated by friction or stored in Leyden jars. These sources could produce only brief discharges, making it difficult to conduct sustained experiments or to harness electricity for practical purposes.

Volta's invention was a direct result of his "contact theory" of electricity, which he had developed in opposition to Luigi Galvani's concept of "animal electricity." Galvani believed that electricity was a vital force inherent in living organisms, while Volta argued that it was a physical phenomenon that could be generated by the contact between dissimilar metals. The voltaic pile was the culmination of Volta's efforts to prove his theory, and its success provided strong evidence for his ideas.

The basic design of the voltaic pile was remarkably simple. It consisted of a stack of alternating discs of two different metals, typically zinc and silver, separated by pieces of cardboard or cloth soaked in a salt solution or a dilute acid. This arrangement was based on Volta's observation that a small electrical potential, or voltage, developed when two different metals were brought into contact, especially when a moist conductor was present between them.

In a typical voltaic pile, a disc of zinc would be placed at the bottom, followed by a disc of silver. On top of the silver disc, a piece of cardboard or cloth soaked in a salt solution would be placed. This sequence – zinc, silver, electrolyte – would be repeated multiple times, creating a stack of alternating layers. The number of layers determined the voltage of the pile, with each "sandwich" of zinc, silver, and electrolyte adding to the overall potential difference.

To use the pile, wires were connected to the top and bottom discs. When these wires were brought into contact, or connected to a circuit, a continuous flow of electric current was produced. This was unlike anything that had been achieved before. The current from a voltaic pile was not a sudden discharge, like that from a Leyden jar, but a steady and sustained flow that could be maintained for a considerable period.

The key to the voltaic pile's operation was the chemical reaction between the metals and the electrolyte. Volta's contact theory, though not entirely accurate by modern standards, correctly identified the importance of the dissimilar metals. We now know that the different electrochemical potentials of the metals cause a flow of electrons when they are connected through an electrolyte.

In the case of a zinc-silver pile, zinc has a greater tendency to lose electrons than silver. When a zinc disc and a silver disc are placed in contact with an electrolyte, the zinc tends to dissolve into the solution, releasing zinc ions and leaving behind electrons on the zinc disc. These electrons then flow through the external circuit to the silver disc, where they combine with positive ions from the solution.

This process creates a potential difference between the two metals, with the zinc disc becoming negatively charged and the silver disc becoming positively charged. The electrolyte serves as a conductor, allowing the flow of ions to complete the circuit. The chemical reactions continue as long as there is zinc to dissolve and ions in the solution to react with the electrons at the silver disc.

The voltage produced by a single pair of zinc and silver discs is relatively small, about 0.76 volts. However, by stacking multiple pairs in series, Volta was able to generate much higher voltages. For example, a pile with 20 pairs of discs could produce a voltage of around 15 volts, enough to produce a noticeable shock or to power simple electrical devices.

Volta's invention was not just a scientific curiosity; it was a practical tool that opened up new possibilities for experimentation and innovation. The ability to generate a continuous flow of electric current allowed scientists to study the properties of electricity in a way that had never been possible before. They could investigate the effects of electricity on various materials, explore the relationship between electricity and magnetism, and even begin to develop practical applications for this new form of energy.

One of the first applications of the voltaic pile was in the field of electrochemistry, the study of the relationship between electricity and chemical reactions. Soon after Volta announced his invention, scientists began using the voltaic pile to decompose water into its constituent elements, hydrogen and oxygen. This process, known as electrolysis, was a groundbreaking discovery that demonstrated the close connection between electricity and chemical change.

In 1800, just weeks after Volta's letter to the Royal Society was made public, two English scientists, William Nicholson and Anthony Carlisle, constructed a voltaic pile and used it to perform the first electrolysis of water. They connected the wires from their pile to two platinum electrodes immersed in a container of water. To their amazement, they observed bubbles of gas forming at each electrode. They collected the gases and found that the gas produced at the negative electrode was hydrogen, while the gas produced at the positive electrode was oxygen.

This experiment was a landmark achievement. It not only provided further evidence for Volta's theory of electricity but also demonstrated that water was not an element, as had been previously believed, but a compound made up of two distinct elements. This discovery had profound implications for chemistry, paving the way for a deeper understanding of the composition of matter and the nature of chemical reactions.

The voltaic pile also played a crucial role in the discovery of new elements. In 1807, the English chemist Humphry Davy used a powerful voltaic pile, consisting of over 250 pairs of plates, to isolate several new elements, including potassium, sodium, calcium, strontium, barium, and magnesium. He accomplished this by passing an electric current through molten salts of these elements, a process that separated the metal from the other components of the salt.

Davy's experiments were a tour de force of electrochemical investigation. He demonstrated that electricity could be used to break down compounds that had previously been considered indivisible. His work not only expanded the known periodic table of elements but also provided further evidence for the close relationship between electricity and chemical forces. He even hypothesized that chemical bonding itself was electrical in nature, an idea that would be further developed in the following centuries.

The voltaic pile also enabled scientists to study the physiological effects of electricity in more detail. While Galvani had explored the effects of static electricity on muscle tissue, the voltaic pile provided a means to apply a continuous current to living organisms. This led to a greater understanding of how nerves and muscles function and how electricity can be used to stimulate or inhibit their activity.

Beyond the realm of scientific research, the voltaic pile also sparked the imagination of inventors and entrepreneurs who saw the potential for using electricity in practical applications. One of the most promising areas was the development of electric lighting. Although it would take several decades to develop a commercially viable electric light bulb, early experiments with the voltaic pile demonstrated the possibility of using electricity to produce light.

Humphry Davy, in addition to his work on electrolysis, also conducted experiments on electric arcs. In 1802, he demonstrated that an intense spark, or arc, could be produced by passing an electric current through two carbon rods that were slightly separated. This arc produced a brilliant light, far brighter than any candle or oil lamp. While Davy's arc lamp was not practical for everyday use, it was the first demonstration of electric lighting and hinted at the possibilities that lay ahead.

The voltaic pile, despite its groundbreaking impact, had several limitations. It was bulky and heavy, making it difficult to transport or use in portable devices. The liquid electrolyte tended to dry out or leak, requiring frequent maintenance. Moreover, the zinc plates were gradually consumed by the chemical reactions, limiting the lifespan of the pile.

These limitations spurred efforts to improve the design of the battery and to develop more practical sources of electric current. One of the first improvements was the invention of the "cruickshank trough" battery by William Cruickshank, an English chemist, around 1802. Instead of stacking the metal discs vertically, Cruickshank arranged them horizontally in a wooden trough, with each pair of plates soldered together. The spaces between the plates were filled with a salt solution or dilute acid. This design was less prone to leakage and easier to assemble than Volta's original pile.

Another important development was the invention of the "Daniell cell" by John Frederic Daniell, an English chemist, in 1836. The Daniell cell used a copper electrode immersed in a copper sulfate solution and a zinc electrode immersed in a zinc sulfate solution. The two solutions were separated by a porous barrier that allowed ions to pass through but prevented the solutions from mixing. This design provided a more stable and long-lasting source of current than the voltaic pile, and it became widely used in telegraphy and other applications.

Despite these improvements, the fundamental principles behind the voltaic pile remained central to battery technology for many decades. The basic concept of using two different metals and an electrolyte to generate electricity through chemical reactions is still used in modern batteries, although the materials and designs have become much more sophisticated.

The invention of the voltaic pile was a pivotal moment in the history of science and technology. It provided the first practical source of continuous electric current, opening up new avenues of research and innovation. It enabled scientists to study the properties of electricity in unprecedented detail, leading to groundbreaking discoveries in electrochemistry, electromagnetism, and physiology. It also sparked the imagination of inventors, setting in motion the development of electric lighting, telegraphy, and countless other technologies that would transform the world.

Volta's legacy extends far beyond his invention of the battery. His work helped to establish electricity as a fundamental force of nature, governed by the same laws that applied to other physical phenomena. His contact theory, though later refined, laid the groundwork for our modern understanding of electrochemistry and the relationship between electricity and chemical reactions. His meticulous experimental approach and his willingness to challenge established ideas set a standard for scientific inquiry that continues to inspire researchers today.

The story of the voltaic pile is a testament to the power of scientific curiosity and the transformative potential of new technologies. It's a story that began with a simple observation about the twitching of frog legs and culminated in a device that would change the world forever. The voltaic pile was not just a scientific breakthrough; it was a turning point in human history, the dawn of a new era of electrical technology that continues to unfold to this day. The humble stack of metal discs and electrolyte-soaked cardboard may seem primitive by today's standards, but it was the spark that ignited the electrical revolution, a revolution that has illuminated our world and continues to shape our lives in countless ways.

CHAPTER EIGHT: Oersted's Discovery: The Intimate Link Between Electricity and Magnetism

The voltaic pile had unleashed a torrent of scientific investigation into the nature of electricity. With a steady source of electric current now available, researchers across Europe were probing the properties of this mysterious force, exploring its effects on various substances, and uncovering its connection to chemical reactions. Yet, despite the rapid progress being made in understanding electricity, its relationship to another equally enigmatic force – magnetism – remained shrouded in uncertainty. For centuries, these two phenomena had been studied separately, their connection, if any, a matter of speculation and conjecture. This was all about to change, thanks to a serendipitous observation made by a Danish physicist named Hans Christian Ørsted.

Ørsted, a professor of physics at the University of Copenhagen, was a prominent figure in the scientific community of his time. He was deeply interested in the philosophical ideas of Immanuel Kant, particularly Kant's concept of the "unity of forces," the idea that all natural forces were interconnected and ultimately derived from a single, fundamental force. This philosophical perspective shaped Ørsted's scientific pursuits, leading him to search for connections between seemingly disparate phenomena, such as electricity, magnetism, heat, and light.

In the winter of 1819, Ørsted was conducting a series of experiments on electricity and magnetism. He had long suspected that there might be a link between the two forces, but he had not yet found any conclusive evidence. He had experimented with magnets near wires carrying an electric current, hoping to see some interaction, but without success. He had even placed the wires at different angles to no avail.

As fate would have it, the crucial breakthrough came during a lecture demonstration in April 1820. Ørsted was demonstrating the heating effect of an electric current by passing a current from a voltaic pile through a thin platinum wire. As he was showing this to his students, he happened to have a compass nearby, a standard piece of equipment used to demonstrate the Earth's magnetic field.

The exact details of what happened next are somewhat unclear, as Ørsted's own accounts are brief and somewhat ambiguous. However, the gist of the story is that he noticed something unexpected: when he switched on the current in the wire, the compass needle, which had been pointing north as usual, deflected and moved to a position almost perpendicular to the wire. When he switched off the current, the needle returned to its original north-pointing position.

This was a momentous observation. For the first time, a direct link between electricity and magnetism had been demonstrated. The movement of the compass needle, a magnetic object, was clearly influenced by the electric current flowing through the wire. This meant that electricity could produce a magnetic effect, a finding that had profound implications for the understanding of both forces.

Ørsted was initially surprised and perplexed by this unexpected result. He had not been specifically looking for this effect during the lecture, and he had previously performed similar experiments without observing any interaction between the current-carrying wire and the compass. It seems that the key difference in this particular instance was the orientation of the wire relative to the compass needle In his earlier experiments, he had placed the wire perpendicular to the needle, and it was only when the wire was parallel to the needle that the effect became apparent.

After the lecture, Ørsted conducted further experiments to investigate this new phenomenon in more detail. He found that the deflection of the compass needle depended on the direction of the current in the wire. When he reversed the current, the needle deflected in the opposite direction. He also observed that the effect was stronger when the current was stronger and when the compass was closer to the wire.

Ørsted's experiments showed that the electric current created a circular magnetic field around the wire. The compass needle, being a small magnet itself, aligned itself with this field, just as it normally aligned itself with the Earth's magnetic field. This was a crucial insight, as it revealed that the magnetic effect of an electric current was not a simple attraction or repulsion, as might have been expected, but a more complex circular force.

Ørsted published his findings in a short paper titled "Experimenta circa effectum conflictus electrici in acum magneticam" (Experiments on the effect of electric conflict on the magnetic needle), which was circulated in Latin to scientists across Europe in July 1820. The paper was only four pages long, but it described his experiments and observations in clear and concise terms, leaving no doubt about the significance of his discovery.

The news of Ørsted's discovery spread rapidly throughout the scientific community, creating a wave of excitement and stimulating a flurry of new research. Scientists across Europe quickly replicated his experiments and began to explore the implications of this newfound connection between electricity and magnetism. The discovery was not just a scientific curiosity; it was a fundamental shift in the understanding of the natural world. It challenged the long-held belief that electricity and magnetism were separate and unrelated forces, and it opened up a whole new field of inquiry that would eventually lead to the development of electromagnetism, one of the cornerstones of modern physics.

One of the first to recognize the importance of Ørsted's work was the French physicist André-Marie Ampère, whose contributions will be discussed in detail in the next chapter. Ampère, upon hearing of Ørsted's discovery, immediately began his own experiments and within weeks had developed a more comprehensive understanding of the relationship between electricity and magnetism. He showed that two parallel wires carrying electric currents attracted each other if the currents were in the same direction and repelled each other if the currents were in opposite directions. He also formulated a mathematical law, known as Ampère's Law, that described the magnetic force between two electric currents.

Ørsted's discovery also spurred the development of new instruments for measuring electric current. The first such instrument, called a "galvanometer," was invented by the German physicist Johann Schweigger soon after Ørsted's publication. Schweigger's galvanometer consisted of a coil of wire with a compass needle suspended inside. When an electric current flowed through the coil, it created a magnetic field that deflected the needle, and the amount of deflection was proportional to the strength of the current. This device allowed scientists to measure electric currents with much greater precision than had been previously possible, and it became an essential tool for electrical research.

The galvanometer, in various forms, would become a fundamental instrument in electrical science and engineering. It was used to measure not only the strength of electric currents but also other electrical quantities, such as voltage and resistance. It played a crucial role in the development of telegraphy, telephony, and other electrical technologies.

Ørsted's discovery also had a profound impact on the philosophical understanding of the relationship between different forces of nature. His work provided strong evidence for the Kantian idea of the "unity of forces," the notion that all natural forces were interconnected and ultimately derived from a single source. This idea would become a guiding principle for many scientists in the 19th century, leading them to search for further connections between electricity, magnetism, light, heat, and other phenomena.

The quest to unify the forces of nature reached its pinnacle in the work of James Clerk Maxwell, whose equations, published in the 1860s, unified electricity, magnetism, and light into a single elegant framework. Maxwell's theory of electromagnetism, which will be discussed in a later chapter, was one of the greatest achievements of 19th-century physics, and it owed a great debt to Ørsted's initial discovery of the magnetic effect of electric current.

While Ørsted's discovery was a major breakthrough, it also raised new questions. Why did an electric current produce a magnetic field? What was the nature of this field? How was it related to the electric current that produced it? These questions would occupy scientists for many decades, leading to a deeper understanding of the fundamental nature of electricity and magnetism.

Ørsted's own research interests extended beyond electromagnetism. He continued to teach and conduct research at the University of Copenhagen, where he made important contributions to the study of acoustics, optics, and the compressibility of liquids and gases. He was also a dedicated educator and played a key role in promoting science education in Denmark. He founded the Society for the Dissemination of Natural Science, an organization dedicated to popularizing science among the general public, and he was instrumental in the establishment of the Polytechnic Institute in Copenhagen, which became a leading center for technical education and research.

Ørsted's legacy extends far beyond his scientific discoveries. He was a key figure in the development of science in Denmark and a tireless advocate for the importance of science education. He was also a respected public intellectual, known for his philosophical writings and his engagement with the broader cultural and intellectual currents of his time. He was a true polymath, with interests that ranged from physics and chemistry to philosophy, literature, and aesthetics.

Ørsted's serendipitous discovery during a lecture demonstration in 1820 was a turning point in the history of science. It revealed the intimate link between electricity and magnetism, two forces that had previously been considered separate and distinct. This discovery opened up a whole new field of research, leading to the development of electromagnetism as a fundamental branch of physics. It also spurred the invention of new instruments and technologies, such as the galvanometer, that would transform the study and application of electricity.

Ørsted's work was not just a scientific breakthrough; it was a philosophical triumph. It provided strong evidence for the interconnectedness of natural forces, a concept that would inspire generations of scientists to seek a deeper understanding of the underlying unity of the universe. His legacy continues to inspire scientists today, as they probe the mysteries of the cosmos and seek to unravel the fundamental laws that govern the behavior of matter and energy. The simple act of noticing a compass needle deflect near a current-carrying wire set in motion a chain of discoveries that would transform our understanding of the world and pave the way for the technological marvels of the modern era.

CHAPTER NINE: Ampère's Insights: Quantifying the Magnetic Force of Current

The news of Ørsted's discovery that an electric current could deflect a compass needle reverberated through the scientific community like a thunderclap. It was a revelation, a clear demonstration that electricity and magnetism, two forces that had long been considered distinct, were intimately connected. Among those who were profoundly impacted by Ørsted's findings was André-Marie Ampère, a French physicist and mathematician. Ampère, upon learning of Ørsted's experiment, immediately recognized its significance and embarked on a series of experiments that would not only elucidate the nature of this newfound relationship but also lay the foundation for the mathematical theory of electromagnetism.

Ampère was a remarkable figure, a self-taught polymath who had made significant contributions to mathematics, chemistry, and physics. He was born in 1775 near Lyon, France, into a well-to-do family. His father, a successful businessman, encouraged his son's intellectual curiosity and provided him with a broad education. Ampère showed an early aptitude for mathematics, devouring advanced treatises on the subject while still in his teens. He was largely self-educated, mastering Latin and Greek to access the works of the great mathematicians and scientists of the past.

The French Revolution dramatically altered the course of Ampère's life. His father, who had served as a local official, was arrested and executed by the Jacobins in 1793. This tragic event deeply affected Ampère, plunging him into a period of grief and depression. He eventually recovered and resumed his studies, taking up a position as a mathematics teacher in Lyon.

In 1801, Ampère moved to Bourg-en-Bresse, where he took up another teaching post. It was during this time that he began his first serious research in mathematics, focusing on the theory of probability and its application to games of chance. His work on the mathematical theory of games earned him recognition and led to his appointment as a professor of mathematics at the prestigious École Polytechnique in Paris in 1809.

At the École Polytechnique, Ampère's research interests broadened to include other areas of physics and chemistry. He was particularly interested in the work of Humphry Davy, the English chemist who had used the voltaic pile to isolate several new elements. Ampère corresponded with Davy and other leading scientists of the day, exchanging ideas and keeping abreast of the latest developments in the rapidly evolving field of electrical science.

It was against this backdrop of intellectual ferment that Ampère learned of Ørsted's discovery in the summer of 1820. He immediately grasped the importance of Ørsted's observation and saw it as a key to unlocking the mysteries of both electricity and magnetism. Within a week of hearing about Ørsted's experiment, Ampère had replicated it himself and begun his own investigations.

Ampère's approach to the problem was characteristically mathematical. He was not content with simply observing the qualitative effects of electric currents on magnets; he wanted to quantify the forces involved, to develop a mathematical law that could describe the interaction between electricity and magnetism with precision.

He began by conducting a series of meticulous experiments, building upon Ørsted's initial observations. He used a variety of experimental setups, carefully measuring the deflection of compass needles under different conditions. He varied the strength and direction of the current, the distance between the wire and the compass, and the orientation of the wire.

One of Ampère's first insights was that the magnetic effect of an electric current was not limited to its influence on compass needles. He realized that if an electric current could exert a force on a magnet, then, by Newton's third law of motion (for every action, there is an equal and opposite reaction), a magnet should also exert a force on a current-carrying wire.

To test this hypothesis, Ampère devised an ingenious experiment. He suspended a section of wire so that it was free to move, and then he brought a magnet close to it. As he passed a current through the wire, he observed that the wire moved, either towards or away from the magnet, depending on the direction of the current and the orientation of the magnet's poles. This was a crucial confirmation of his hypothesis: a magnet did indeed exert a force on a current-carrying wire.

Ampère then took his investigation a step further. If a magnet could exert a force on a current-carrying wire, and if, as Ørsted had shown, an electric current produced a magnetic effect, then it followed that two current-carrying wires should exert forces on each other. This was a bold prediction, and Ampère set out to test it.

He constructed an apparatus with two parallel wires, each suspended so that it was free to move. When he passed currents through the wires, he observed that they either attracted or repelled each other, depending on the direction of the currents. If the currents flowed in the same direction, the wires attracted each other; if the currents flowed in opposite directions, the wires repelled each other.

This was a groundbreaking discovery. Ampère had demonstrated that electric currents interacted with each other through magnetic forces, even in the absence of any permanent magnets. This meant that magnetism was not an inherent property of certain materials, like iron, but a fundamental aspect of electricity in motion.

Ampère's experiments were not limited to parallel wires. He explored the interaction between wires at different angles and in different configurations. He showed that the force between two current-carrying wires was proportional to the product of the currents and inversely proportional to the distance between them.

He also found that the force depended on the relative orientation of the wires.

Through these meticulous experiments, Ampère developed a comprehensive understanding of the magnetic forces between electric currents. He expressed his findings in a mathematical formula, now known as Ampère's force law, which described the force between two current elements, small segments of current-carrying wires.

Ampère's force law was a remarkable achievement. It was the first quantitative description of the relationship between electricity and magnetism, and it provided a foundation for the development of a unified theory of electromagnetism. It allowed scientists to calculate the magnetic forces between electric currents in various configurations, and it provided a powerful tool for designing and analyzing electrical devices.

Ampère's work went beyond simply quantifying the forces between currents. He also developed a theoretical framework for understanding the nature of magnetism itself. He proposed that all magnetic phenomena were ultimately due to electric currents, even the magnetism of permanent magnets like lodestones.

He hypothesized that within a permanent magnet, there were tiny, circular currents of electricity flowing at the molecular level. He called these "molecular currents" or "Amperian currents." He suggested that in magnetic materials, these molecular currents were aligned in the same direction, creating a macroscopic magnetic field. In non-magnetic materials, the molecular currents were randomly oriented, canceling each other out.

This was a radical idea at the time, as it suggested that magnetism was not a separate entity but a manifestation of electricity in motion. It also provided an explanation for why magnets always had two poles, north and south. If a magnet was cut in half, each half would still have both a north and a south pole because the molecular currents would still be circulating in the same direction within each piece.

Ampère's theory of molecular currents was remarkably prescient. Although the details of his model were not entirely accurate by modern standards, the fundamental concept that magnetism is due to circulating electric currents is a cornerstone of our current understanding of magnetism. We now know that the magnetic properties of materials are due to the intrinsic spin of electrons and their orbital motion around atoms, which create tiny magnetic dipole moments, the equivalent of Ampere's molecular currents.

Ampère presented his findings in a series of papers published in the early 1820s, culminating in his magnum opus, "Mémoire sur la théorie mathématique des phénomènes électrodynamiques, uniquement déduite de l'expérience" (Memoir on the Mathematical Theory of Electrodynamic Phenomena, Uniquely Deduced from Experiment), published in 1827.

In this work, Ampère laid out his comprehensive theory of electrodynamics, as he called the study of the interaction between electric currents. He presented his force law, his theory of molecular currents, and a wealth of experimental evidence to support his ideas. He also developed a mathematical framework for calculating the magnetic field produced by electric currents in various configurations.

One of the key mathematical tools that Ampère developed was a law that related the integral of the magnetic field around a closed loop to the total current passing through the loop. This law, now known as Ampère's circuital law, is one of the fundamental equations of electromagnetism. It's a powerful tool for calculating magnetic fields in situations with high degrees of symmetry, such as around a long, straight wire or inside a solenoid (a coil of wire).

Ampère's circuital law is analogous to Gauss's law in electrostatics, which relates the electric flux through a closed surface to the total charge enclosed within the surface. Together, these two laws form an important part of the foundation of classical electromagnetism, providing a way to relate electric and magnetic fields to their sources.

Ampère's work was not immediately accepted by all of his contemporaries. Some scientists found his mathematical approach difficult to follow, and others were skeptical of his theory of molecular currents. However, as the evidence in favor of his ideas mounted, and as his mathematical framework proved its power in solving practical problems, his work gradually gained acceptance.

The impact of Ampère's work on the development of electrical science was profound. He not only established the quantitative relationship between electricity and magnetism but also provided a theoretical framework for understanding the nature of magnetism itself. His work laid the foundation for the development of electromagnetism as a unified field of study, and it paved the way for the technological innovations that would transform the world in the 19th and 20th centuries.

Ampère's insights led directly to the development of more sophisticated instruments for measuring and manipulating electric currents. The galvanometer, which had been invented shortly after Ørsted's discovery, was refined and improved, thanks in part to Ampère's work. The ability to measure currents accurately was crucial for the development of telegraphy, electric motors, and other electrical technologies.

Ampère's legacy extends beyond his specific scientific contributions. He is remembered as one of the pioneers of the scientific method, a meticulous experimenter who combined careful observation with rigorous mathematical analysis. His work exemplifies the power of combining experimental investigation with theoretical modeling, an approach that remains central to scientific progress today.

The unit of electric current, the ampere (often shortened to "amp"), is named in his honor. This is a fitting tribute to a man who dedicated his life to understanding the nature of electric currents and their relationship to magnetism.

Ampère's story is a testament to the power of scientific curiosity, the importance of rigorous experimentation, and the profound impact that a single individual can have on the course of scientific history. His work transformed our understanding of electricity and magnetism, laying the groundwork for the technological revolution that would reshape the world in ways he could scarcely have imagined. He was a true visionary, a scientist whose insights illuminated the hidden connections between the forces of nature and whose legacy continues to inspire scientists and engineers to this day. His insights were not merely incremental advances but represented a fundamental shift in the understanding of the physical world, demonstrating that seemingly disparate phenomena could be understood as different manifestations of the same underlying principles.

CHAPTER TEN: Faraday's Breakthrough: Electromagnetic Induction

The scientific world was still buzzing from the discoveries of Ørsted and Ampère, who had revealed the profound connection between electricity and magnetism. Electric currents, it was now clear, could create magnetic fields, and these fields could exert forces on other currents and magnets. This was a revolutionary shift in understanding, but it also raised a fundamental question: if electricity could produce magnetism, could magnetism, in turn, produce electricity?

This question captivated the mind of Michael Faraday, a self-taught English scientist working at the Royal Institution in London. Faraday, a man of humble origins and limited formal education, would rise to become one of the most influential experimental physicists in history. His relentless curiosity, his remarkable experimental skills, and his intuitive grasp of physical phenomena would lead him to a series of groundbreaking discoveries that would transform our understanding of electricity and magnetism and pave the way for the development of much of the electrical technology we rely on today.

Faraday was born in 1791 in Newington Butts, a village near London. His father was a blacksmith, and his family struggled with poverty. Faraday's formal education was limited to basic reading, writing, and arithmetic. At the age of 14, he was apprenticed to a bookbinder, a trade that would have a profound impact on his life.

The seven years Faraday spent as a bookbinder's apprentice were not just a period of manual labor; they were also an opportunity for intellectual growth. Faraday took advantage of his access to books, reading voraciously on a wide range of subjects, including science. He was particularly fascinated by articles on electricity in the "Encyclopædia Britannica."

Inspired by what he read, Faraday began to conduct his own simple experiments, building Leyden jars and other electrical apparatus from materials he could find. He also attended lectures at the City Philosophical Society, where he learned about the latest scientific discoveries and theories.

Faraday's life took a dramatic turn in 1812 when he attended a series of lectures given by the renowned chemist Humphry Davy at the Royal Institution. Davy, famous for his work on electrolysis and his discovery of several new elements, was a charismatic speaker and a leading figure in the scientific world.

Faraday was captivated by Davy's lectures. He took meticulous notes, carefully bound them into a book, and sent them to Davy as a token of his admiration and a request for employment. Davy was impressed by Faraday's dedication and enthusiasm, and in 1813, he hired the young man as a laboratory assistant at the Royal Institution.

Faraday's early years at the Royal Institution were spent assisting Davy in his chemical experiments. He learned the techniques of chemical analysis, and he assisted Davy in his work on the safety lamp for miners, an invention that saved countless lives by preventing explosions in coal mines.

However, Faraday's interests increasingly turned towards electricity and magnetism. He was particularly intrigued by the work of Ørsted and Ampère, and he began to conduct his own experiments on the interaction between electric currents and magnetic fields.

In 1821, Faraday made his first major discovery in this field. He constructed a device in which a wire carrying an electric current rotated around a fixed magnet. This was the first demonstration of what is now known as electromagnetic rotation, and it was the precursor to the electric motor.

Faraday's device, though simple, was a significant breakthrough. It showed that the magnetic field produced by an electric current could be used to create continuous motion. This was a crucial step towards harnessing electricity for practical applications, and it hinted at the possibility of converting electrical energy into mechanical work.

However, Faraday was not content with simply demonstrating electromagnetic rotation. He was driven by a deeper question: if electricity could produce magnetism, could magnetism produce electricity? He believed that there must be a symmetry between the two forces, a reciprocal relationship that had not yet been discovered.

Faraday began a long and arduous search for this elusive connection. He conducted numerous experiments, wrapping coils of wire around magnets, moving magnets near wires, and trying various other configurations, but all to no avail. He could not find any evidence that magnetism, by itself, could produce an electric current.

Despite these initial failures, Faraday did not give up. He meticulously documented his experiments in his laboratory notebooks, filling thousands of pages with detailed descriptions of his setups, his observations, and his thoughts. He was a master of experimental design, always seeking new ways to probe the relationship between electricity and magnetism.

Faraday's persistence finally paid off in 1831. In a series of groundbreaking experiments, he discovered the phenomenon of electromagnetic induction, the ability of a changing magnetic field to induce an electric current in a nearby conductor. This was the breakthrough he had been seeking, the missing piece of the puzzle that completed the picture of the relationship between electricity and magnetism.

Faraday's first successful experiment involved a simple apparatus consisting of two coils of wire wound around an iron ring. One coil, which he called the primary coil, was connected to a battery. The other coil, the secondary coil, was connected to a galvanometer, an instrument used to detect electric current.

Faraday reasoned that if a changing magnetic field could induce an electric current, then the act of switching the current on or off in the primary coil should create a momentary change in the magnetic field around the iron ring. This changing magnetic field, he hypothesized, should then induce a current in the secondary coil.

To test this idea, Faraday connected the primary coil to the battery and watched the galvanometer connected to the secondary coil. To his astonishment, he saw the galvanometer needle deflect momentarily when he connected the battery and again when he disconnected it. This was the effect he had been searching for: a changing magnetic field was indeed inducing an electric current in a nearby conductor.

Faraday's initial observation was just the beginning. He conducted a series of further experiments, varying the conditions and exploring the nature of this newly discovered phenomenon. He found that the induced current in the secondary coil only flowed when the current in the primary coil was changing, either when it was switched on or off. A steady current in the primary coil produced no current in the secondary coil.

This was a crucial observation. It showed that it was not magnetism itself that induced the current, but a *change* in the magnetic field. This was a fundamental insight that would have profound implications for the understanding of electromagnetism.

Faraday continued his experiments, exploring different ways to create a changing magnetic field. He discovered that he could induce a current in a coil of wire by simply moving a magnet in and out of the coil. The faster he moved the magnet, the stronger the induced current. He also found that he could induce a current by rotating a coil of wire in a magnetic field.

Through these experiments, Faraday established the basic principles of electromagnetic induction. He showed that a changing magnetic field could induce an electric current in a nearby conductor, and that the magnitude of the induced current was proportional to the rate of change of the magnetic field.

Faraday's discovery of electromagnetic induction was a watershed moment in the history of science. It was the counterpart to Ørsted's discovery of the magnetic effect of electric current, completing the picture of the intimate relationship between electricity and magnetism. It showed that these two forces were not just interconnected but were, in fact, two aspects of the same fundamental force: electromagnetism.

Faraday's work on induction was not just a theoretical breakthrough; it also had profound practical implications. It laid the foundation for the development of the electric generator, a device that converts mechanical energy into electrical energy. The generator, in its various forms, would become the cornerstone of the electrical power industry, enabling the widespread generation and distribution of electricity that would transform the world in the late 19th and 20th centuries.

Faraday, with his remarkable intuition for physical phenomena, recognized the practical potential of his discovery. He demonstrated a simple generator by rotating a copper disc between the poles of a horseshoe magnet. This device, known as the Faraday disc, was the first rudimentary electric generator, capable of producing a continuous, albeit small, electric current.

Faraday's discovery of electromagnetic induction also provided a crucial piece of the puzzle that would eventually lead to a unified theory of electromagnetism. James Clerk Maxwell, building upon Faraday's work, would later develop his famous equations that described the behavior of electric and magnetic fields and their interaction with each other. Maxwell's equations, a cornerstone of classical physics, showed that light itself was an electromagnetic wave, a profound insight that unified electricity, magnetism, and optics into a single theoretical framework.

Faraday's experimental approach to science was as important as his discoveries themselves. He was a master of experimental design, meticulously documenting his work in his laboratory notebooks. He had a remarkable ability to visualize physical phenomena, and he often used analogies and models to understand complex concepts.

One of Faraday's most famous conceptual tools was the idea of "lines of force." He visualized electric and magnetic fields as consisting of lines that represented the direction and strength of the field at each point in space. These lines of force were not just a mathematical abstraction; Faraday believed that they were real physical entities, a kind of stress or strain in the space surrounding electric charges and magnets.

The concept of lines of force was a powerful tool for understanding the behavior of electric and magnetic fields. It allowed Faraday to visualize how these fields interacted with each other and how they could be used to induce currents or create motion. This concept was later developed mathematically by Maxwell, who incorporated it into his theory of electromagnetism.

Faraday's work on electricity and magnetism earned him widespread recognition and numerous honors. He was elected a Fellow of the Royal Society in 1824, and he received the Royal Medal and the Copley Medal, the Society's highest awards. He was also offered the presidency of the Royal Society but declined, preferring to focus on his research.

Despite his fame and success, Faraday remained a humble and unassuming man throughout his life. He was deeply religious, a member of the Sandemanian Church, a small Christian sect that emphasized simple living and adherence to biblical principles. His faith played an important role in his life and work, providing him with a sense of purpose and a framework for understanding the natural world.

Faraday continued to work at the Royal Institution until his retirement in 1861. He died on August 25, 1867, at the age of 75. His legacy, however, lived on, continuing to inspire generations of scientists and engineers.

Faraday's contributions to science were not limited to his work on electricity and magnetism. He made important discoveries in chemistry, including the discovery of benzene and the laws of electrolysis. He also conducted research on optics, acoustics, and other areas of physics.

However, it is his work on electromagnetism that stands as his greatest achievement. His discovery of electromagnetic induction was a pivotal moment in the history of science, revealing the deep connection between electricity and magnetism and paving the way for the development of technologies that would transform the world. His experimental approach, his intuitive grasp of physical phenomena, and his concept of lines of force were instrumental in shaping our modern understanding of electromagnetism.

Faraday's story is a testament to the power of curiosity, perseverance, and the importance of experimental investigation. He was a self-taught scientist who rose from humble beginnings to become one of the most influential figures in the history of physics. His work not only transformed our understanding of the natural world but also laid the foundation for the electrical age, an era that continues to unfold to this day.

CHAPTER ELEVEN: The Electric Motor: From Scientific Curiosity to Practical Application

The first ten chapters of our history have followed a path of profound discovery, a journey driven by insatiable curiosity and the unyielding quest to understand the fundamental forces of nature. From the ancient Greeks' fascination with amber to Faraday's groundbreaking discovery of electromagnetic induction, the pursuit of knowledge surrounding electricity and magnetism was largely confined to the realm of philosophical inquiry and scientific experimentation. While these discoveries had illuminated the intimate relationship between electricity and magnetism, their practical applications remained, for the most part, a tantalizing vision of the future.

The electric motor, however, would change all that. This remarkable invention, born from the seeds of scientific curiosity, would blossom into a powerful force that would transform industries, revolutionize transportation, and ultimately reshape the very fabric of modern life. The journey from abstract scientific principles to a practical, working electric motor was not a single, instantaneous leap but rather a gradual, iterative process, fueled by the ingenuity and perseverance of numerous inventors and scientists across Europe and America.

The foundations of the electric motor can be traced back to Faraday's demonstration of electromagnetic rotation in 1821. His simple device, in which a current-carrying wire rotated around a fixed magnet, was the first demonstration that electrical energy could be converted into continuous mechanical motion. This was a pivotal moment, a proof of concept that hinted at the vast potential of harnessing electricity to perform useful work.

However, Faraday's device was far from a practical motor. It produced only a small amount of rotational force and was primarily a scientific curiosity. The challenge facing inventors was to take this basic principle and transform it into a device capable of delivering substantial power and performing meaningful tasks.

One of the earliest attempts to build a practical electric motor was made by William Sturgeon, an English physicist and inventor, in 1832. Sturgeon, inspired by the work of both Ørsted and Faraday, had already made significant contributions to the field of electromagnetism. He had invented the electromagnet in 1825, a device that consisted of a coil of wire wound around an iron core. When an electric current passed through the coil, the iron core became magnetized, and when the current was switched off, the magnetism disappeared. This ability to create and control a magnetic field with electricity was a crucial step towards the development of the electric motor.

Sturgeon's motor consisted of a wooden wheel with several electromagnets mounted around its rim. A commutator, a device that reversed the direction of the electric current in the electromagnets at the appropriate moment, was used to create a continuous rotation. As the wheel turned, the commutator switched the current in the electromagnets, causing them to attract and repel a set of fixed permanent magnets, thus producing a continuous rotational force.

Sturgeon's motor was a significant improvement over Faraday's earlier device. It was capable of producing more power and could run continuously for extended periods. However, it was still relatively weak and inefficient, and its primary use was for demonstrating the principles of electromagnetism rather than for performing practical work.

Around the same time, across the Atlantic, an American blacksmith and inventor named Thomas Davenport was independently working on his own version of the electric motor. Davenport, a self-taught man with a keen interest in science and technology, had learned about electromagnetism from a demonstration of Joseph Henry's powerful electromagnets. Henry, an American scientist who had independently discovered electromagnetic induction around the same time as Faraday, had built some of the most powerful electromagnets of his day.

Inspired by Henry's work, Davenport set out to build an electric motor. He purchased one of Henry's electromagnets and, through trial and error, developed his own motor design. In 1834, he built a small, battery-powered motor that used two electromagnets, one fixed and one rotating, to create continuous motion.

Davenport's motor was a remarkable achievement, especially considering his lack of formal scientific training. He recognized the potential of his invention and, unlike many inventors of his time, he sought patent protection for his motor. In 1837, he was granted one of the first patents for an electric motor in the United States.

Davenport was not just an inventor; he was also an entrepreneur. He envisioned a wide range of applications for his motor, from powering tools and machinery to driving vehicles. He built several larger motors, including one that he used to power a small electric car that ran on a circular track. This was one of the first demonstrations of electric-powered transportation, a concept that would revolutionize travel in the years to come.

Davenport also used his motor to power a printing press, demonstrating its potential for industrial applications. He even established a workshop in New York City to manufacture and sell his motors. Despite his efforts, however, Davenport's motors were ultimately not commercially successful. They were expensive to produce, and the batteries of the time were not powerful or long-lasting enough to make them a viable alternative to existing power sources like steam engines.

Despite the commercial failure of his motors, Davenport's work was significant. He demonstrated the feasibility of building practical electric motors and showcased their potential for a variety of applications. He also helped to popularize the idea of electric power, sparking the imagination of other inventors and entrepreneurs.

While Sturgeon and Davenport were making strides in the development of practical electric motors, other inventors were also working on similar projects. In Germany, Moritz Hermann Jacobi, a Prussian-born architect and physicist, built an electric motor in 1834 that he used to power a small boat on the Neva River in St. Petersburg, Russia. Jacobi's motor was relatively powerful for its time, and his demonstration of electric-powered water transportation was a significant milestone.

Jacobi continued to refine his motor designs, and in 1838, he built a larger, more powerful motor that propelled a 28-foot boat carrying 14 passengers. This demonstration garnered considerable attention and helped to further establish the potential of electric motors for transportation.

Despite these early successes, the electric motor remained a relatively niche technology for several decades. The limitations of battery technology, the high cost of producing motors, and the lack of a widespread electrical infrastructure hindered their widespread adoption. Steam engines remained the dominant source of power for industry and transportation throughout much of the 19th century.

However, the development of the electric motor continued, driven by the ongoing improvements in battery technology and the growing understanding of electromagnetism. Inventors and scientists continued to experiment with different motor designs, seeking to improve their efficiency, power, and reliability.

One important development was the invention of the Gramme dynamo by Zénobe Gramme, a Belgian electrical engineer, in 1870. The Gramme dynamo was a type of direct current (DC) generator that was significantly more efficient and powerful than previous generators. It was also capable of being used as a motor, a feature that made it particularly versatile.

The Gramme dynamo was a major breakthrough, providing a more practical and reliable source of electric power for motors and other electrical devices. It was quickly adopted for a variety of applications, including powering arc lights, driving machinery in factories, and even powering early electric trams.

The availability of more powerful and efficient generators, like the Gramme dynamo, spurred further innovation in motor design. Inventors began to develop motors specifically designed to run on direct current, taking advantage of the characteristics of this type of power source.

One of the key figures in the development of DC motors was Frank J. Sprague, an American electrical engineer and inventor. Sprague, a former naval officer who had worked with Thomas Edison, recognized the potential of electric motors for transportation. In 1887, he developed a practical DC motor and control system that he used to power an electric streetcar system in Richmond, Virginia.

Sprague's streetcar system was a resounding success. It was the first large-scale demonstration of electric-powered urban transportation, and it proved that electric streetcars were a viable and efficient alternative to horse-drawn carriages and cable cars. Sprague's system was quickly adopted by other cities, and within a few years, electric streetcars were becoming a common sight in urban areas around the world.

The success of electric streetcars spurred further development of DC motors and control systems. Sprague continued to refine his designs, and other inventors and engineers also contributed to the advancement of this technology. DC motors became increasingly powerful, efficient, and reliable, and they found applications in a wide range of industries, from powering elevators and cranes to driving machinery in factories and mines.

The development of the electric motor was not limited to DC designs. In the late 19th century, a new type of motor, based on alternating current (AC), began to emerge. AC motors, which will be discussed in more detail in later chapters, offered several advantages over DC motors, including the ability to operate at higher voltages and to transmit power more efficiently over long distances.

The development of both DC and AC motors was a crucial step in the electrification of the world. These motors provided a clean, efficient, and versatile source of power that could be used in a wide range of applications. They transformed industries, revolutionized transportation, and brought electric power into homes and businesses, forever changing the way people lived and worked.

The journey from Faraday's simple demonstration of electromagnetic rotation to the powerful and sophisticated electric motors of the late 19th century was a long and winding one. It involved the contributions of numerous inventors, scientists, and engineers, each building upon the work of those who came before them. It was a process of trial and error, of experimentation and refinement, driven by the pursuit of a common goal: to harness the power of electricity to perform useful work.

The electric motor, once a scientific curiosity, had become a driving force of progress. It was a testament to the power of human ingenuity and the transformative potential of scientific discovery. The story of the electric motor is not just a story about technology; it's a story about how scientific breakthroughs can be translated into practical applications that improve people's lives and reshape the world around us. It's a story that continues to unfold today, as electric motors continue to evolve and find new applications in areas such as electric vehicles, robotics, and renewable energy. The legacy of those early pioneers, from Faraday to Davenport to Sprague, lives on in the hum of every electric motor, a reminder of the remarkable journey from scientific curiosity to practical application.

CHAPTER TWELVE: The Generator: Harnessing Mechanical Energy to Produce Electricity

The electric motor, as we saw in the previous chapter, marked a pivotal moment in the history of electricity. It demonstrated that electrical energy could be converted into mechanical work, opening up a world of possibilities for powering machinery, transportation, and countless other applications. However, the widespread adoption of electric motors was initially hampered by the limitations of the primary source of electricity at the time: batteries. While batteries were suitable for small-scale applications and laboratory experiments, they were expensive, bulky, and not capable of producing the large amounts of power needed to drive industrial machinery or light entire cities.

The solution to this problem lay in the very principle that made the electric motor possible: electromagnetic induction. Faraday's groundbreaking discovery in 1831 had revealed that a changing magnetic field could induce an electric current in a nearby conductor. This meant that not only could electricity produce motion, as demonstrated by the motor, but motion could also produce electricity. This was the key to unlocking the full potential of electrical power, and it led to the development of the electric generator, a device that would transform the world in ways that even Faraday could scarcely have imagined.

The electric generator, in its simplest form, is essentially an electric motor working in reverse. Instead of using electricity to produce motion, it uses motion to produce electricity. The basic principle is to move a conductor, such as a coil of wire, through a magnetic field. This movement creates a changing magnetic flux through the conductor, which, according to Faraday's law of induction, induces an electromotive force (EMF), or voltage, in the conductor. This voltage can then drive a current through an external circuit, thus producing electrical power.

While Faraday had demonstrated the basic principle of electromagnetic induction and even built a simple generator, the Faraday disc, it was others who would take this concept and develop it into practical, large-scale generators capable of producing substantial amounts of electricity. The journey from Faraday's initial discovery to the powerful generators that would eventually power the modern world was a complex one, involving the contributions of numerous inventors, engineers, and scientists across Europe and America.

One of the first significant steps towards a practical generator was taken by Hippolyte Pixii, a French instrument maker, in 1832, just a year after Faraday's discovery. Pixii's generator consisted of a horseshoe magnet that was rotated by a hand crank. As the magnet rotated, its poles passed by a coil of wire wound around an iron core. This created a changing magnetic field in the coil, inducing an alternating current (AC) that flowed first in one direction and then in the other.

Pixii's generator was a significant improvement over Faraday's disc, as it produced a much stronger current. However, the alternating current it produced was not suitable for all applications. Many early electrical devices, including some types of motors and electroplating systems, required direct current (DC), which flows in only one direction.

To address this issue, Pixii added a commutator to his generator. The commutator, which had been invented earlier for use in electric motors, was a clever device that reversed the connections to the coil every half turn, effectively converting the alternating current into a pulsating direct current. While this pulsating DC was not as smooth as the current from a battery, it was suitable for many applications and represented a major step towards a practical generator.

Pixii's generator, though a significant advance, was still relatively inefficient and produced a weak current. It was primarily used for scientific demonstrations and laboratory experiments rather than for practical applications. However, it inspired other inventors to improve upon its design, leading to a series of innovations that gradually increased the power and efficiency of generators.

One such innovation was the use of multiple coils and magnets. By increasing the number of coils and magnets, inventors found that they could significantly increase the output of the generator. This was because each coil contributed to the total induced voltage, and each magnet contributed to the changing magnetic field.

Another important development was the invention of the magneto-electric generator, or "magneto," by E.M. Clarke, an English instrument maker, in 1835. Clarke's magneto used powerful permanent magnets to create the magnetic field, eliminating the need for a separate source of electricity to power electromagnets, as was the case in some earlier designs. This made the magneto more self-contained and easier to use.

Magnetos became widely used for a variety of applications, including powering telegraph systems, detonating explosives, and providing electric shocks for medical purposes. They were also used in early telephones, where they generated the ringing current.

While magnetos were an improvement over earlier generators, they still had limitations. The strength of the magnetic field was limited by the strength of the permanent magnets, and the output was still relatively low. The next major breakthrough came with the development of generators that used electromagnets to create the magnetic field, a concept known as "self-excitation."

The idea of using electromagnets in a generator was not new. Several inventors had experimented with this concept, but they had typically used a separate source of electricity, such as a battery, to power the electromagnets. This made the generator less practical and more complicated.

The key innovation of self-excitation was to use a portion of the generator's own output to power the electromagnets. This created a feedback loop, where the stronger the magnetic field, the more current was induced, and the more current was induced, the stronger the magnetic field became. This allowed the generator to build up a powerful magnetic field from a small initial current, greatly increasing its output.

The principle of self-excitation was independently discovered by several inventors in the 1850s and 1860s. One of the first was Søren Hjorth, a Danish inventor, who patented a self-excited generator in 1854. However, Hjorth's design was not widely publicized, and it did not have an immediate impact on the development of generator technology.

Another key figure in the development of self-excitation was Ányos Jedlik, a Hungarian Benedictine priest, physicist, and inventor. Jedlik built a self-excited generator, which he called an "electromagnetic self-rotor," in 1861, but he did not publish his work or seek a patent, and his invention remained largely unknown outside of Hungary for many years.

The concept of self-excitation gained wider recognition through the work of several other inventors, including Sir Charles Wheatstone, a prominent English scientist and inventor, and Werner von Siemens, a German electrical engineer and industrialist. Both Wheatstone and Siemens independently developed self-excited generators in 1866 and presented their work to the scientific community.

Siemens, in particular, played a crucial role in promoting the development and adoption of self-excited generators. He recognized the commercial potential of this technology and formed a company, Siemens & Halske, to manufacture and sell generators and other electrical equipment. Siemens' generators, which he called "dynamo-electric machines," or simply "dynamos," were quickly adopted for a variety of applications, including arc lighting, electroplating, and powering industrial machinery.

The development of the dynamo was a major turning point in the history of electricity. It provided a practical and efficient means of generating large amounts of electrical power, paving the way for the widespread adoption of electric lighting, electric motors, and other electrical technologies. The dynamo was also a key component of the first central power stations, which began to appear in the 1880s, providing electricity to entire cities.

One of the most important improvements to the dynamo was made by Zénobe Gramme, a Belgian electrical engineer, in 1870. Gramme's dynamo used a ring armature, a continuous winding of wire around an iron ring, which produced a much smoother and more powerful direct current than previous designs. The Gramme dynamo was also capable of being used as a motor, a feature that made it particularly versatile.

The Gramme dynamo was a major commercial success, and it was quickly adopted for a variety of applications. It was used to power arc lights in lighthouses, theaters, and public squares. It was also used to drive machinery in factories and to power early electric trams. The Gramme dynamo helped to establish the practicality and reliability of electric power, and it played a crucial role in the growth of the electrical industry.

The development of the generator continued throughout the late 19th and early 20th centuries. Engineers and inventors continued to refine dynamo designs, improving their efficiency, power output, and reliability. They experimented with different configurations of magnets and coils, different types of commutators, and different materials.

One important development was the invention of the drum armature by Friedrich von Hefner-Alteneck, an engineer working for Siemens, in 1872. The drum armature, which consisted of coils wound longitudinally around a cylindrical core, was more efficient and produced a higher voltage than the ring armature. It quickly became the standard design for DC generators.

Another significant advance was the development of multipolar generators, which used multiple pairs of magnetic poles to increase the output voltage and reduce the speed at which the generator needed to rotate. This made generators more compact and efficient, and it allowed them to be directly coupled to steam engines and other prime movers.

The late 19th century also saw the rise of alternating current (AC) generators, also known as alternators. While DC generators were well-suited for many applications, they had limitations when it came to transmitting power over long distances. AC generators, which produced a current that alternated in direction, offered several advantages for power transmission, as will be discussed in later chapters.

The development of practical AC generators was a complex process, involving the contributions of numerous inventors and engineers. One of the key figures in this field was Nikola Tesla, a Serbian-American inventor and electrical engineer, who developed a polyphase AC induction motor and a corresponding system of alternators, transformers, and transmission lines. Tesla's system, which was adopted by the Westinghouse Electric Company, played a crucial role in the "War of the Currents" between AC and DC, ultimately leading to the triumph of AC as the dominant standard for power generation and distribution.

The invention and development of the electric generator was a pivotal moment in the history of technology. It was the key that unlocked the full potential of electricity, transforming it from a scientific curiosity into a powerful force that would reshape the world. The generator made it possible to produce large amounts of electricity efficiently and reliably, paving the way for the widespread adoption of electric lighting, electric motors, and countless other electrical technologies that we take for granted today.

The development of the generator was not a single event but a long and complex process, involving the contributions of many individuals across different countries and over several decades. It was a story of ingenuity, perseverance, and collaboration, as inventors, engineers, and scientists built upon each other's work, gradually improving the design and performance of generators.

From Pixii's simple hand-cranked generator to the powerful dynamos of Gramme and Siemens, and from the early magnetos to the sophisticated alternators of Tesla, the evolution of the generator was a testament to the power of human innovation. It was a journey driven by the desire to harness the power of nature for the benefit of humanity, a journey that would ultimately lead to the electrification of the world and the dawn of a new era of technological progress. The hum of generators in power plants around the world is a constant reminder of the remarkable achievement of those early pioneers who transformed Faraday's fundamental discovery into a practical technology that powers our modern world. The generator stands as a symbol of the transformative power of scientific discovery and its ability to shape the course of human history.

CHAPTER THIRTEEN: Maxwell's Equations: Unifying Electricity, Magnetism, and Light

The latter half of the 19th century witnessed a period of intense intellectual ferment in the field of physics. The discoveries of Ørsted, Ampère, and Faraday had revealed the profound interconnectedness of electricity and magnetism, sparking a revolution in scientific thought. Yet, despite the remarkable progress that had been made, a unified theory of electromagnetism remained elusive. This was the challenge that James Clerk Maxwell, a brilliant Scottish physicist, set out to tackle. Maxwell, armed with a formidable mathematical mind and a deep intuition for physical phenomena, would not only synthesize the existing knowledge of electricity and magnetism but also make a breathtaking prediction that would transform our understanding of light and lay the foundation for much of modern physics.

Maxwell was born in Edinburgh in 1831, into a family of comfortable means and intellectual distinction. His father, a lawyer, recognized his son's early aptitude for learning and encouraged his intellectual pursuits. Maxwell showed an early interest in geometry and mechanics, constructing intricate polyhedra and other geometric models. He received a rigorous education at the Edinburgh Academy and later at the University of Edinburgh, where he studied mathematics, physics, and philosophy.

In 1850, Maxwell moved to the University of Cambridge, the epicenter of British science at the time. At Cambridge, he immersed himself in the study of mathematics and physics, excelling in his studies and earning a reputation as a brilliant and original thinker. He was particularly influenced by the work of William Thomson (later Lord Kelvin), a leading physicist who had made significant contributions to the fields of thermodynamics and electromagnetism.

After graduating from Cambridge, Maxwell held academic positions at Marischal College in Aberdeen and later at King's College London. It was during these years that he began to focus his research on electricity and magnetism, building upon the experimental work of Faraday.

Maxwell was deeply impressed by Faraday's concept of "lines of force," the idea that electric and magnetic fields were not just mathematical abstractions but real physical entities that permeated space. Faraday, though lacking in formal mathematical training, had a remarkable ability to visualize physical phenomena, and his lines of force provided a powerful conceptual tool for understanding the behavior of electric and magnetic fields.

Maxwell, with his strong mathematical background, set out to translate Faraday's intuitive ideas into a rigorous mathematical framework. He believed that a mathematical theory of electromagnetism was not just possible but necessary to fully understand the relationship between electricity, magnetism, and the mysterious lines of force that Faraday had envisioned.

Maxwell's first major work on electromagnetism, "On Faraday's Lines of Force," was published in two parts between 1855 and 1856. In this paper, he developed a mathematical model of Faraday's lines of force, using the analogy of an incompressible fluid flowing through space. He showed that the behavior of electric and magnetic fields could be described by a set of equations that governed the flow of this imaginary fluid.

While Maxwell's fluid model was ultimately a mechanical analogy, it was a crucial step towards a mathematical theory of electromagnetism. It allowed him to express Faraday's ideas in a precise mathematical language, and it provided a framework for further development.

Maxwell's next major contribution came in his 1861-1862 paper, "On Physical Lines of Force." In this work, he abandoned the fluid analogy and developed a more abstract model based on the concept of a "molecular vortex" field. He imagined space as being filled with tiny, rotating cells, or vortices, that represented the magnetic field. The rotation of these vortices, he proposed, was responsible for the magnetic force.

In this model, Maxwell introduced the concept of "displacement current," a crucial addition to Ampère's law. Ampère's law, as originally formulated, related the magnetic field around a closed loop to the electric current passing through the loop. However, Maxwell realized that this law was incomplete, as it did not account for the effects of changing electric fields.

Maxwell's displacement current was a revolutionary idea. He proposed that a changing electric field could produce a magnetic field, just as a changing magnetic field could produce an electric field, as Faraday had shown. This concept completed the symmetry between electricity and magnetism, and it was essential for a consistent mathematical theory of electromagnetism.

To illustrate the concept of displacement current, Maxwell used the example of a charging capacitor. A capacitor consists of two conductive plates separated by an insulator. When a capacitor is connected to a battery, current flows into one plate and out of the other, creating a buildup of charge on the plates.

According to Ampère's law, there should be no magnetic field between the plates of the capacitor, as there is no actual flow of electric current through the insulating gap. However, Maxwell argued that the changing electric field between the plates as the capacitor charges is equivalent to a current, which he called the displacement current. This displacement current, he proposed, produces a magnetic field just like a real current.

The introduction of displacement current was not just a mathematical trick; it had profound physical implications. It meant that changing electric and magnetic fields were inextricably linked, each capable of producing the other. This concept was the key to understanding the propagation of electromagnetic waves, as Maxwell would soon demonstrate.

In his 1865 paper, "A Dynamical Theory of the Electromagnetic Field," Maxwell presented his most comprehensive and influential work on electromagnetism. In this paper, he brought together his previous work, refined his mathematical model, and presented a set of equations that described the behavior of electric and magnetic fields in their full generality.

These equations, now known as Maxwell's equations, are considered one of the greatest achievements of 19th-century physics. They are a set of four coupled partial differential equations that describe how electric and magnetic fields are generated and interact with each other and with matter.

Here are Maxwell's equations in their modern vector calculus form:

1. Gauss's law for electricity: $\nabla \cdot \mathbf{E} = \rho/\varepsilon_0$
2. Gauss's law for magnetism: $\nabla \cdot \mathbf{B} = 0$
3. Faraday's law of induction: $\nabla \times \mathbf{E} = -\partial \mathbf{B}/\partial t$
4. Ampère-Maxwell's law: $\nabla \times \mathbf{B} = \mu_0(\mathbf{J} + \varepsilon_0 \, \partial \mathbf{E}/\partial t)$

Where:

- **E** is the electric field
- **B** is the magnetic field
- ρ is the electric charge density
- **J** is the electric current density
- ε_0 is the permittivity of free space
- μ_0 is the permeability of free space
- $\nabla \cdot$ is the divergence operator
- $\nabla \times$ is the curl operator

- ∂/∂t is the partial derivative with respect to time

These equations, though compact in their modern form, encapsulate a wealth of information about the behavior of electric and magnetic fields. Let's briefly examine each equation:

1. **Gauss's law for electricity** states that the electric flux through a closed surface is proportional to the total electric charge enclosed within the surface. In simpler terms, it describes how electric charges create electric fields.

2. **Gauss's law for magnetism** states that the magnetic flux through a closed surface is always zero. This implies that there are no isolated magnetic poles, or magnetic monopoles, unlike electric charges. Magnetic field lines always form closed loops.

3. **Faraday's law of induction** describes how a changing magnetic field creates an electric field. This is the principle behind electric generators.

4. **Ampère-Maxwell's law** describes how both electric currents and changing electric fields create magnetic fields. The term $\varepsilon_0 \, \partial E/\partial t$ represents Maxwell's displacement current, the crucial addition that completed the symmetry between electricity and magnetism.

Maxwell's equations were a monumental achievement. They unified all known phenomena of electricity and magnetism into a single, elegant theoretical framework. They showed that electricity and magnetism were not separate forces but two aspects of the same fundamental force: the electromagnetic force.

But Maxwell's work went beyond simply unifying electricity and magnetism. His equations had a startling implication, one that Maxwell himself recognized and that would revolutionize our understanding of light

By manipulating his equations, Maxwell showed that they allowed for the existence of self-propagating electromagnetic waves. These waves consisted of oscillating electric and magnetic fields, perpendicular to each other and to the direction of propagation. The changing electric field produced a magnetic field, and the changing magnetic field produced an electric field, sustaining the wave as it traveled through space.

Even more remarkably, Maxwell calculated the speed of these electromagnetic waves and found that it was equal to the speed of light, a value that had been previously measured through astronomical observations and terrestrial experiments. This was an astonishing result. The speed of light, a fundamental constant of nature, emerged naturally from Maxwell's equations, which were based on the laws of electricity and magnetism.

Maxwell realized the profound implication of this result: light itself was an electromagnetic wave. This was a revolutionary idea, as it unified three seemingly disparate phenomena – electricity, magnetism, and light – into a single theoretical framework. Light, which had been studied for centuries as a separate phenomenon, was now understood to be fundamentally electromagnetic in nature.

In his 1865 paper, Maxwell boldly stated: "This velocity is so nearly that of light, that it seems we have strong reason to conclude that light itself (including radiant heat, and other radiations if any) is an electromagnetic disturbance in the form of waves propagated through the electromagnetic field according to electromagnetic laws."

Maxwell's prediction that light was an electromagnetic wave was a triumph of theoretical physics. It was a bold and far-reaching conclusion, deduced from a set of equations that described the behavior of electric and magnetic fields. It was a testament to the power of mathematical reasoning and its ability to reveal profound truths about the natural world.

Maxwell's theory also predicted the existence of electromagnetic waves beyond the visible spectrum of light. He suggested that there could be electromagnetic waves with wavelengths longer or shorter than those of visible light, invisible to the human eye but still fundamentally electromagnetic in nature.

This prediction opened up a whole new realm of scientific inquiry. It suggested that the electromagnetic spectrum was far broader than previously thought, encompassing a vast range of wavelengths, from radio waves at the long-wavelength end to gamma rays at the short-wavelength end, with visible light occupying only a small portion of this spectrum.

The experimental verification of Maxwell's theory would have to wait for several years, as the technology needed to generate and detect electromagnetic waves beyond the visible spectrum did not yet exist. It would be another 23 years until Heinrich Hertz, in 1888, would provide the first definitive experimental evidence for the existence of radio waves, confirming Maxwell's prediction and ushering in the era of wireless communication.

Maxwell's work had a profound and lasting impact on physics. His equations became one of the cornerstones of classical physics, alongside Newton's laws of motion and the laws of thermodynamics. They provided a complete and accurate description of the behavior of electric and magnetic fields at the macroscopic level, and they laid the foundation for much of modern technology, from electric power generation and distribution to telecommunications and medical imaging.

Maxwell's equations also played a crucial role in the development of Einstein's theory of special relativity. Einstein was deeply impressed by the symmetry and elegance of Maxwell's theory, and he recognized that it was incompatible with Newtonian mechanics. In particular, Maxwell's equations predicted that the speed of light was a universal constant, independent of the motion of the observer, a result that contradicted the Newtonian concept of relative velocities.

Einstein's quest to reconcile Maxwell's theory with the principles of mechanics led him to develop the theory of special relativity, which revolutionized our understanding of space, time, and the nature of light. Special relativity showed that Maxwell's equations were, in fact, more fundamental than Newton's laws, and that Newtonian mechanics was only an approximation that held true at low speeds.

Maxwell's legacy extends far beyond his equations. He made significant contributions to other areas of physics, including thermodynamics, statistical mechanics, and the kinetic theory of gases. He was a pioneer in the use of statistical methods in physics, and his work on the distribution of molecular velocities in a gas laid the foundation for the field of statistical mechanics.

Maxwell was also a gifted teacher and communicator of science. He wrote several influential textbooks, including "Theory of Heat" and "Treatise on Electricity and Magnetism," which helped to disseminate his ideas and to train a new generation of physicists.

Despite his relatively short life – he died of abdominal cancer at the age of 48 – Maxwell's impact on science was immense. His work transformed our understanding of electricity, magnetism, and light, and it laid the foundation for much of modern physics and technology. His equations remain as relevant and powerful today as they were when he first formulated them, a testament to their fundamental importance and their enduring beauty. Maxwell's profound insights into the nature of the electromagnetic field continue to illuminate our understanding of the universe, from the smallest subatomic particles to the largest cosmic structures. His legacy is a shining example of the power of human intellect to unlock the secrets of nature and to reveal the hidden unity that underlies the seemingly diverse phenomena of the physical world.

CHAPTER FOURTEEN: The Dawn of Electric Lighting: Edison and the Incandescent Bulb

The late 19th century was a time of rapid technological advancement, fueled by the burgeoning understanding of electricity and its potential applications. The development of practical generators had made it possible to produce electricity on a large scale, and the electric motor had demonstrated the power of electricity to perform useful work. Yet, one of the most transformative applications of electricity, one that would forever change the way people lived and worked, was still in its nascent stages: electric lighting.

For centuries, humanity had relied on fire as its primary source of artificial light. Candles, oil lamps, and gaslights provided illumination, but they were often dim, flickering, and required constant maintenance. They also produced smoke, heat, and unpleasant odors, making them far from ideal for indoor use. The advent of electricity offered the tantalizing possibility of a cleaner, brighter, and more convenient form of lighting, and many inventors and entrepreneurs were eager to capitalize on this potential.

The story of electric lighting is not the tale of a single invention or a lone genius, but rather a complex narrative of incremental improvements, competing technologies, and fierce commercial rivalries. It's a story that involves numerous individuals across Europe and America, each contributing a piece to the puzzle, gradually transforming the dream of electric light into a reality.

One of the earliest attempts to create electric light was made by Humphry Davy, the English chemist who had employed a young Michael Faraday as his assistant, in the early 19th century. In 1802, Davy demonstrated that an electric current could heat a thin strip of platinum to incandescence, producing a bright, white light. This was the first demonstration of incandescent lighting, a principle that would eventually become the dominant form of electric lighting.

Davy also experimented with electric arcs, a phenomenon he first observed in 1808. An electric arc is a luminous discharge that occurs when a high voltage is applied across a gap between two electrodes. The intense heat of the arc vaporizes the electrode material, creating a brilliant, sustained light. Davy's arc lamp, which used two carbon rods as electrodes, was the first practical demonstration of electric arc lighting.

While Davy's experiments were groundbreaking, they were far from practical for widespread use. His incandescent platinum strips burned out quickly, and his arc lamp required a large and expensive battery to operate. Furthermore, the arc lamp produced an intensely bright and harsh light that was unsuitable for indoor use.

Despite these limitations, Davy's work sparked the imagination of other inventors, who began to explore the possibilities of electric lighting. Throughout the first half of the 19th century, numerous attempts were made to improve upon Davy's designs, but progress was slow. The lack of a practical and affordable source of electricity remained a major obstacle.

The development of the dynamo in the 1860s and 1870s provided a crucial breakthrough. With a reliable source of electricity now available, inventors could focus on developing practical electric lights. Two main approaches emerged: arc lighting and incandescent lighting.

Arc lighting, based on Davy's earlier work, was the first to achieve commercial success. Arc lamps, which used two carbon rods as electrodes, produced a brilliant, intense light that was well-suited for outdoor illumination. In the 1870s, arc lighting systems began to be installed in streets, squares, and lighthouses in major cities across Europe and America.

One of the key figures in the development of arc lighting was Pavel Yablochkov, a Russian electrical engineer. In 1876, Yablochkov invented the "Yablochkov candle," an improved arc lamp that used two parallel carbon rods separated by an insulating material. The Yablochkov candle was simpler and more reliable than previous arc lamps, and it could be easily replaced when the carbon rods burned down.

Yablochkov's candles were widely adopted for street lighting in Paris, London, and other cities. They were often installed in groups, mounted on tall poles, creating a dramatic and impressive display of electric illumination. However, arc lighting had several drawbacks. The light was extremely bright and harsh, making it unsuitable for indoor use. The carbon rods needed to be replaced frequently, and the lamps produced flickering, noise, and unpleasant odors.

While arc lighting was gaining popularity for outdoor illumination, many inventors were pursuing a different approach: incandescent lighting. The goal was to create a lamp that produced a softer, more pleasant light suitable for indoor use. The basic principle was to heat a thin filament of material to incandescence by passing an electric current through it.

The challenge was to find a suitable filament material that could withstand the high temperatures required for incandescence without burning out too quickly. Early attempts used platinum, carbonized paper, and other materials, but none proved to be both practical and long-lasting.

One of the early pioneers of incandescent lighting was Warren de la Rue, an English astronomer and chemist. In 1840, he designed an incandescent lamp that used a coiled platinum filament enclosed in a vacuum tube. The vacuum helped to prevent the filament from oxidizing and burning out too quickly. However, de la Rue's lamp was expensive to produce and did not last very long.

Another important figure in the development of incandescent lighting was Joseph Swan, an English physicist and chemist. Swan began experimenting with incandescent lamps in the 1850s, using carbonized paper filaments. He faced the same challenges as other inventors: finding a durable filament and creating a suitable vacuum to prevent oxidation.

Swan's work was interrupted for several years, but he returned to it in the 1870s, spurred by the improvements in vacuum pump technology. By 1878, he had developed a working incandescent lamp that used a treated cotton thread as a filament. Swan's lamp was enclosed in a glass bulb that had been evacuated of air, creating a partial vacuum.

Swan demonstrated his lamp publicly in Newcastle, England, in 1878, and he began installing his lighting system in homes and businesses. He also established a company, the Swan Electric Light Company, to manufacture and sell his lamps. Swan's work was a major breakthrough, and he is often credited as the co-inventor of the incandescent light bulb.

However, across the Atlantic, another inventor was also making significant progress in the development of incandescent lighting: Thomas Edison. Edison, an American inventor and entrepreneur, would become synonymous with the electric light bulb, and his name would forever be associated with this transformative invention.

Edison was a prolific inventor, with over a thousand patents to his name. He was also a shrewd businessman and a master of self-promotion. He established the first industrial research laboratory in Menlo Park, New Jersey, where he assembled a team of talented engineers and technicians to work on a variety of projects, including the telegraph, the phonograph, and, of course, the electric light.

Edison recognized the enormous commercial potential of electric lighting, and he set out to develop a complete system of electric illumination, including not just the lamp itself but also the generators, distribution networks, and other components necessary to make electric lighting a practical and affordable alternative to gaslight.

Edison and his team at Menlo Park began experimenting with incandescent lamps in 1878. They tested thousands of different materials for the filament, including platinum, carbonized paper, and various types of plant fibers. They also experimented with different bulb shapes, vacuum levels, and methods of sealing the bulb.

Edison's approach was systematic and relentless. He famously said, "I have not failed. I've just found 10,000 ways that won't work." His team worked day and night, meticulously testing and refining their designs.

In October 1879, Edison and his team achieved a major breakthrough. They tested a lamp with a filament made from carbonized cotton thread, and it burned for over 13 hours. This was a significant improvement over previous attempts, and it demonstrated that a practical incandescent lamp was within reach.

Edison continued to refine his lamp design, experimenting with different types of carbon filaments and improving the vacuum inside the bulb. He also developed a screw-in base for the lamp, which became the standard for incandescent bulbs and is still used today.

Edison's most famous demonstration of his electric lighting system took place on New Year's Eve, 1879. He illuminated his Menlo Park laboratory and the surrounding streets with dozens of his incandescent lamps, creating a spectacular display that attracted thousands of visitors. The event was widely publicized, and it cemented Edison's reputation as the inventor of the electric light.

Edison's success was not just due to his inventive genius but also to his business acumen. He recognized that the key to the widespread adoption of electric lighting was to create a complete system, not just a better lamp. He established the Edison Electric Light Company to manufacture and sell his lamps, and he began building central power stations to generate and distribute electricity.

The first commercial central power station, Pearl Street Station, opened in New York City in 1882. It used coal-fired steam engines to drive direct current (DC) generators, which supplied electricity to a network of underground cables. The station initially served a small area of lower Manhattan, providing power to about 85 customers with a total of 400 lamps.

The success of Pearl Street Station marked the beginning of the electric age. Edison's system of electric lighting quickly spread to other cities, and his company, which eventually became General Electric, grew into a major corporation. Other companies also entered the electric lighting market, including the Westinghouse Electric Company, which championed the use of alternating current (AC) for power transmission.

The competition between Edison's DC system and Westinghouse's AC system, known as the "War of the Currents," was a fierce and often bitter rivalry. Edison, who had invested heavily in DC technology, argued that AC was dangerous and unsafe. He even went so far as to publicly electrocute animals using AC to demonstrate its supposed dangers.

Despite Edison's efforts, AC ultimately proved to be more practical and efficient for long-distance power transmission. AC could be easily stepped up to high voltages using transformers, which reduced transmission losses, and then stepped back down to lower voltages for use in homes and businesses. DC, on the other hand, could not be easily transformed, and it required thicker, more expensive copper wires for transmission.

By the early 1890s, AC had become the dominant standard for power generation and distribution, and Edison's DC system was gradually phased out. However, Edison's contributions to the development of electric lighting and the creation of a complete electrical system were undeniable. He played a crucial role in making electric light a practical reality and in bringing the benefits of electricity to the masses.

The invention of the incandescent light bulb was not a single event but a gradual process involving the contributions of many individuals. While Edison is often credited as the sole inventor, it's important to recognize the work of other pioneers, such as Swan, who independently developed their own versions of the incandescent lamp.

The development of electric lighting was a transformative moment in human history. It changed the way people lived and worked, extending the hours of productivity and leisure. It made cities safer and more vibrant, and it brought a new level of convenience and comfort to homes and businesses. The electric light bulb, a seemingly simple device, became a symbol of progress and innovation, and it ushered in a new era of technological advancement. It illuminated not just homes and streets but also the path towards a future where electricity would play an ever-increasing role in shaping the human experience. The story of electric lighting is a testament to the power of human ingenuity, the importance of perseverance, and the profound impact that technological innovation can have on society.

CHAPTER FIFTEEN: The Battle of the Currents: AC vs. DC

The late 19th century witnessed a technological clash of titans, a fierce competition that would determine the very foundation of the electrical age. This was the "War of the Currents," a battle fought not with bullets and bombs, but with volts and amps. On one side stood Thomas Edison, the celebrated inventor of the incandescent light bulb and champion of direct current (DC). On the other side were George Westinghouse and Nikola Tesla, proponents of the upstart alternating current (AC) system. At stake was not just the potentially enormous fortune to be made from controlling the future of electric power distribution, but also the very shape of the modern world.

Direct current, as its name suggests, is a type of electrical current that flows in only one direction. It's the kind of electricity produced by batteries and was the first type of current to be used in commercial power systems. Edison, with his vast network of patents and his influential reputation, had built his electrical empire on DC. His Pearl Street Station, the first commercial central power plant in the United States, supplied DC power to a small area of lower Manhattan, illuminating homes and businesses with his incandescent lamps.

Edison's DC system, however, had a significant limitation: it was inefficient at transmitting power over long distances. Due to resistance in the wires, much of the electrical energy was lost as heat, especially when transmitted at the low voltages that were considered safe for use in homes and businesses. This meant that DC power plants needed to be located within a mile or two of their customers, requiring a large number of small, localized power stations to serve a city.

Alternating current, on the other hand, is a type of electrical current that periodically reverses its direction, flowing back and forth in a circuit. The number of times the current changes direction per second is called its frequency, measured in hertz (Hz). The key advantage of AC is that its voltage can be easily transformed using a device called a transformer.

A transformer consists of two coils of wire wound around a common iron core. When an alternating current flows through one coil, it creates a changing magnetic field in the core, which in turn induces an alternating current in the other coil. The ratio of the number of turns in the two coils determines the ratio of the voltages in the two circuits. This means that AC voltage can be stepped up to very high levels for efficient transmission over long distances and then stepped back down to safer levels for use in homes and businesses.

The ability to transform AC voltage was a game-changer. It meant that power plants could be located far from consumers, taking advantage of economies of scale and access to cheaper fuel sources like hydroelectric power. Fewer, larger power plants could serve vast areas, transmitting electricity over high-voltage transmission lines that snaked across the countryside.

The potential advantages of AC were recognized by several inventors and entrepreneurs, but it was George Westinghouse, a successful industrialist who had made his fortune in the railroad industry with his invention of the air brake, who saw the true potential of AC and had the resources to develop it into a viable alternative to Edison's DC system.

Westinghouse, unlike Edison, was not primarily an inventor himself but rather a shrewd businessman and a keen judge of talent. He recognized the limitations of DC and saw in AC the future of electric power. He began acquiring patents related to AC technology and assembling a team of engineers to develop a complete AC system, including generators, transformers, transmission lines, and motors.

One of Westinghouse's most important acquisitions was the patent for an improved AC motor developed by Nikola Tesla, a brilliant and eccentric Serbian-American inventor. Tesla, who had briefly worked for Edison but left after a disagreement over payment and the direction of their research, had independently conceived of a revolutionary new type of motor that used rotating magnetic fields to produce motion.

Tesla's motor, known as the induction motor, was a marvel of engineering. It had no brushes or commutator, making it simpler, more reliable, and more efficient than existing DC motors. It was also perfectly suited for use with AC power, as the alternating current naturally created the rotating magnetic field needed to drive the motor.

Tesla's induction motor was a key component of Westinghouse's AC system. It provided a practical and efficient way to convert AC power into mechanical work, making it suitable for a wide range of industrial and commercial applications. With Tesla's motor and a team of talented engineers, Westinghouse was poised to challenge Edison's dominance in the electric power industry.

The battle between AC and DC, which became known as the "War of the Currents," was not just a technical competition; it was also a public relations war, a struggle for the hearts and minds of the public and the policymakers who would decide the future of the electrical grid. Both sides engaged in a variety of tactics, some legitimate and some questionable, to promote their respective systems and discredit the opposition.

Edison, a master of publicity and self-promotion, launched a vigorous campaign to convince the public that AC was dangerous and unsuitable for widespread use. He and his supporters emphasized the high voltages used in AC transmission, arguing that they posed a serious risk of electrocution. They published pamphlets and articles warning of the dangers of AC, and they even staged public demonstrations in which animals were electrocuted using AC power to highlight its supposed lethality.

One of the most notorious episodes in the War of the Currents was Edison's involvement in the development of the electric chair. In the late 1880s, New York State was looking for a more humane method of execution than hanging, and a commission was formed to investigate alternatives. Edison, seeing an opportunity to further discredit AC, secretly funded the development of an electric chair that used alternating current. He hoped that by associating AC with the death penalty, he could convince the public that it was too dangerous for use in their homes.

The first execution by electric chair, using an AC generator supplied by Westinghouse (against his strong objections), took place in 1890. The execution was botched and gruesome, but Edison and his supporters continued to promote the electric chair as a "humane" alternative to hanging, all while using it as a propaganda tool against AC.

Westinghouse and his allies fought back against Edison's attacks. They argued that AC was perfectly safe when handled properly and that the high voltages used in transmission lines were not a threat to the public. They pointed out that DC systems also used high voltages in some applications, such as streetcar systems, and that the real issue was not the type of current but the safety measures used in its generation, transmission, and distribution.

Westinghouse also emphasized the economic and practical advantages of AC. He argued that AC's ability to transmit power efficiently over long distances would make electricity more affordable and accessible, especially for areas outside of major cities. He also touted the advantages of Tesla's induction motor, which was more efficient and reliable than comparable DC motors.

The War of the Currents was not just fought in the press and in public demonstrations; it was also fought in the marketplace. Westinghouse aggressively bid on contracts to supply electrical systems to cities and businesses, often undercutting Edison's prices. He won several high-profile contracts, including the contract to illuminate the 1893 World's Columbian Exposition in Chicago.

The Chicago World's Fair was a major turning point in the War of the Currents. Westinghouse's AC system powered the fair's lighting, providing a spectacular display of the capabilities of alternating current. Millions of visitors from around the world witnessed the dazzling illumination, and many were impressed by the efficiency and versatility of AC power.

Another major victory for Westinghouse came with the contract to build the hydroelectric power plant at Niagara Falls. This was a landmark project, the first large-scale hydroelectric plant in the world, and it would demonstrate the feasibility of transmitting electricity over long distances using AC.

The power plant, which began operation in 1895, used massive AC generators designed by Tesla to harness the power of the Niagara River. The electricity was transmitted over 20 miles to Buffalo, New York, at a voltage of 11,000 volts, a feat that would have been impossible with DC.

The success of the Niagara Falls project was a major blow to Edison's DC system. It proved that AC was not only safe and reliable but also the most practical and economical way to generate and transmit electricity over long distances. It marked the beginning of the end for DC as the dominant standard for power distribution.

Despite the mounting evidence in favor of AC, Edison continued to fight for his DC system. He invested in improvements to DC technology, hoping to overcome its limitations in long-distance transmission. However, it was a losing battle. The advantages of AC were simply too great to ignore.

By the late 1890s, the War of the Currents was effectively over. AC had emerged as the clear victor, and it became the standard for power generation and distribution around the world. Edison's General Electric eventually abandoned its commitment to DC and began manufacturing AC equipment as well.

The triumph of AC was a pivotal moment in the history of technology. It paved the way for the widespread electrification of the world, bringing electric light, power, and countless other electrical technologies to homes, businesses, and industries across the globe. It made possible the modern electrical grid, a vast and interconnected network of power plants, transmission lines, and distribution systems that is one of the most complex and essential technological achievements of the modern era.

The War of the Currents was not just a battle between two competing technologies; it was a clash between two different visions of the future. Edison, the established giant of the electrical industry, clung to his familiar DC technology, while Westinghouse, the challenger, embraced the innovative potential of AC. It was a contest between a localized, decentralized system and a centralized, interconnected one.

The outcome of the war had far-reaching consequences. The adoption of AC as the standard for power distribution shaped the development of the electrical industry for the next century. It determined the design of power plants, the layout of cities, and the very structure of modern society.

The legacy of the War of the Currents extends beyond the technical details of AC vs. DC. It serves as a case study in the dynamics of technological change, the role of competition in driving innovation, and the importance of infrastructure in shaping the development of new technologies. It also highlights the complex interplay between technology, business, and public perception.

The story of the War of the Currents is a reminder that technological progress is not always a smooth or linear process. It often involves fierce competition, conflicting interests, and public debates about safety, cost, and the direction of technological development. It's a story about how a new technology can disrupt established industries and create new opportunities, and how the choices made during these periods of transition can have long-lasting consequences. The battle between AC and DC was a defining moment in the history of electricity, a turning point that set the stage for the electrical age and transformed the world in ways that its participants could scarcely have imagined. It was a war ultimately won by the inherent superiority of a groundbreaking new method for delivering electricity - one which had clear advantages which could not be overcome by the incumbent technology, or by a smear campaign against it, no matter how well-funded.

CHAPTER SIXTEEN: Tesla's Vision: Alternating Current Triumphs

The War of the Currents, as we saw in the previous chapter, had reached its climax with alternating current (AC) emerging as the clear victor. The advantages of AC for long-distance power transmission were simply too significant to ignore, and even Thomas Edison, the staunch champion of direct current (DC), had been forced to concede defeat. However, the triumph of AC was not solely due to the inherent superiority of the technology itself. It was also the result of the vision, ingenuity, and perseverance of one man: Nikola Tesla.

Tesla, a Serbian-American inventor, engineer, and futurist, was a true visionary, a man whose ideas were often far ahead of his time. He was a pioneer of wireless communication, radio, and X-rays, and he held over 300 patents worldwide. But it was his work on alternating current that would have the most profound and lasting impact on the world. Tesla's understanding of AC, his invention of the induction motor, and his development of a complete polyphase AC system were instrumental in the triumph of alternating current and the electrification of the modern world.

Tesla was born in 1856 in Smiljan, a small village in what is now Croatia, then part of the Austrian Empire. His father was a Serbian Orthodox priest, and his mother was an inventor of household appliances. Tesla showed an early aptitude for science and mathematics, and he was fascinated by electricity from a young age. He studied engineering at the Austrian Polytechnic in Graz, where he first encountered the Gramme dynamo, a DC generator that could also be used as a motor.

Tesla was intrigued by the Gramme dynamo, but he was also troubled by its limitations. The dynamo used a commutator, a device with brushes that reversed the current direction to produce direct current. Tesla saw the commutator as a source of inefficiency and unreliability, as the brushes sparked and wore down over time. He believed that there must be a better way to generate and utilize electricity, a way that did not involve the cumbersome and spark-producing commutator.

While still a student, Tesla conceived of the idea of using alternating current to create a rotating magnetic field that could drive a motor without the need for brushes or a commutator. This was a revolutionary concept, as AC was not widely used at the time, and its properties were not well understood. Tesla's professors dismissed his idea as impossible, a "perpetual motion machine," but he remained convinced that his vision was sound.

After leaving Graz, Tesla worked for several telecommunications companies in Europe, gaining practical experience in electrical engineering. In 1882, while working in Budapest, he had a breakthrough. In a flash of insight, he visualized a complete AC system, including an alternator to generate the current, a motor to convert it into mechanical work, and transformers to step the voltage up and down for efficient transmission.

At the heart of Tesla's system was his induction motor, a revolutionary device that used a rotating magnetic field to induce current in a rotor, causing it to spin. Unlike DC motors, Tesla's induction motor had no physical connection between the stationary part of the motor (the stator) and the rotating part (the rotor). This meant that there were no brushes or commutators to wear out or cause sparking, making the motor simpler, more reliable, and more efficient.

Tesla's induction motor was based on the principle of a rotating magnetic field. He realized that by using multiple alternating currents, each slightly out of phase with the others, he could create a magnetic field that rotated smoothly in space. This rotating field would then induce currents in the rotor, creating a magnetic field that interacted with the rotating field of the stator, causing the rotor to spin.

The concept of a rotating magnetic field was a stroke of genius. It was a completely new way of thinking about electric motors, and it solved the problems that had plagued earlier designs. Tesla's induction motor was not just a theoretical concept; he built working models to demonstrate its feasibility.

In 1884, Tesla emigrated to the United States, hoping to find greater opportunities to develop his ideas. He briefly worked for Thomas Edison in New York City, but the two men had a fundamental disagreement about the future of electric power. Edison was firmly committed to DC, while Tesla was convinced that AC was the superior system.

Tesla left Edison's company and, after a period of struggle, found investors who were willing to back his AC system. In 1887, he established his own laboratory in New York City and began to refine his designs for AC generators, motors, and transformers. He also developed a complete polyphase AC system, which used multiple alternating currents with different phases to transmit power more efficiently.

Tesla's polyphase system was another key innovation. Instead of using a single alternating current, as in earlier AC systems, Tesla's system used two or more currents, each shifted in phase relative to the others. This created a more uniform and powerful rotating magnetic field in the motor, and it also allowed for more efficient power transmission.

Tesla's most common system used two-phase AC, with two currents 90 degrees out of phase with each other. He later developed a three-phase system, with three currents 120 degrees out of phase, which became the standard for power generation and distribution around the world.

In 1888, Tesla received several patents for his AC system, including his induction motor and his polyphase system. He also gave a landmark lecture before the American Institute of Electrical Engineers (AIEE), demonstrating his AC motor and explaining the principles of his system. The lecture caused a sensation in the engineering world, and it brought Tesla to the attention of George Westinghouse.

Westinghouse, as we've seen, was already committed to developing AC technology, but he recognized that Tesla's inventions, particularly his induction motor, were crucial to the success of his system. He negotiated with Tesla to license his patents, paying him a substantial sum and agreeing to pay royalties on each horsepower of AC equipment sold.

The partnership between Tesla and Westinghouse was a turning point in the War of the Currents. With Tesla's patents and Westinghouse's resources, the AC system was now a formidable competitor to Edison's DC system. Westinghouse began to build larger and more powerful AC generators, and he developed transformers that could step up the voltage for long-distance transmission and then step it back down for safe use in homes and businesses.

The advantages of AC, particularly for power transmission, were becoming increasingly apparent. AC could be transmitted at high voltages over long distances with minimal loss, allowing for larger, more efficient power plants to be located far from consumers. DC, on the other hand, required thicker wires and multiple generating stations close to consumers, making it more expensive and less practical for widespread use.

One of the most dramatic demonstrations of the superiority of AC came with the contract to illuminate the 1893 World's Columbian Exposition in Chicago. Westinghouse, using Tesla's AC system, underbid Edison and won the contract. The fair was a spectacular showcase for AC power, with over 200,000 incandescent lamps illuminating the buildings and grounds.

Tesla himself designed the lighting system for the fair, using his polyphase AC generators and transformers. He also displayed his induction motors and other AC devices, demonstrating their versatility and efficiency. Millions of visitors from around the world witnessed the dazzling display of electric light, and many were convinced that AC was the future of electric power.

The success of the Chicago World's Fair was a major blow to Edison's DC system. It showed that AC could be used safely and effectively on a large scale, and it demonstrated the advantages of Tesla's polyphase system. The fair also helped to popularize the use of electricity, creating a growing demand for electric lighting and power.

Another major triumph for AC came with the construction of the hydroelectric power plant at Niagara Falls. This was a project of unprecedented scale, and it would demonstrate the feasibility of transmitting large amounts of power over long distances using AC.

Tesla played a key role in the design of the Niagara Falls power plant. He insisted on using his polyphase AC system, despite opposition from some of the engineers involved in the project. He designed the massive AC generators that would harness the power of the falls, and he oversaw the construction of the transmission lines that would carry the electricity to Buffalo, New York, over 20 miles away.

The Niagara Falls power plant began operation in 1895, and it was an immediate success. It supplied electricity to Buffalo and other nearby cities, powering streetlights, factories, and homes. It was a powerful demonstration of the capabilities of AC power, and it helped to cement AC's position as the dominant standard for power generation and distribution.

The success of the Niagara Falls project marked the end of the War of the Currents. AC had emerged as the clear victor, and even Edison's General Electric began to manufacture AC equipment. The world was rapidly electrifying, and it was Tesla's vision of an AC-powered future that was becoming a reality.

Despite his crucial role in the triumph of AC, Tesla did not reap the full financial rewards of his inventions. In a remarkable act of generosity, he agreed to release Westinghouse from the royalty agreement in their contract. Westinghouse was facing financial difficulties, and Tesla, recognizing the importance of keeping the company afloat to ensure the widespread adoption of AC, tore up the contract that would have made him one of the wealthiest men in the world.

Tesla's decision to forgo his royalties was a testament to his character and his commitment to his vision. He was more interested in seeing his inventions benefit humanity than in amassing personal wealth. He believed that electricity was a force that should be available to all, and he saw AC as the key to making that possible.

After his work with Westinghouse, Tesla continued to invent and innovate for the rest of his life. He established his own laboratory in New York City and pursued a wide range of research, including wireless communication, radio, X-rays, and robotics. He developed the Tesla coil, a high-frequency transformer that is still used today in radio and television sets and for producing high-voltage sparks for scientific experiments and entertainment.

Tesla's later years were marked by financial difficulties and increasing eccentricity. He became obsessed with certain ideas, such as wireless power transmission and the "death ray," a purported particle-beam weapon. He made extravagant claims about his inventions, some of which were never realized or even properly documented. He lived alone in a series of New York hotels, often relying on the generosity of friends and admirers to support him.

Tesla died in 1943 at the age of 86, largely forgotten by the public. However, his legacy has endured, and his contributions to science and technology have been increasingly recognized in recent years. He is now seen as one of the most important inventors and visionaries of the modern era, a man whose ideas were far ahead of his time.

Tesla's impact on the world is undeniable. His AC system revolutionized the way electricity is generated, transmitted, and used. It made possible the modern electrical grid, bringing electric light and power to billions of people around the globe. His induction motor is still the most widely used type of electric motor, powering everything from household appliances to industrial machinery.

Tesla's work also laid the foundation for many other technologies that we take for granted today. His research on high-frequency alternating currents paved the way for the development of radio, television, and other forms of wireless communication. His experiments with high voltages and electric discharges contributed to the development of X-ray technology and neon lighting.

Beyond his specific inventions, Tesla's greatest legacy is perhaps his visionary approach to science and technology. He was a true futurist, a man who could see beyond the limitations of existing technology and imagine a world transformed by new discoveries and inventions. He believed in the power of science to improve the human condition, and he dedicated his life to pursuing that vision.

Tesla's story is a reminder that progress often comes from those who dare to challenge the status quo, who are willing to think differently and to pursue their ideas with passion and determination. It's a story about the power of imagination, the importance of perseverance, and the profound impact that a single individual can have on the course of history. Tesla's vision of an AC-powered world was once considered a radical and impractical dream. Today, it is the reality that powers our homes, our cities, and our lives. His legacy continues to inspire scientists, engineers, and inventors around the world, a testament to the enduring power of his ideas and the brilliance of his mind. His inventions and discoveries were instrumental in bringing about the electrification of the world, and his name will forever be associated with the triumph of alternating current and the dawn of the electrical age. It was Tesla's unique ability to combine a deep theoretical understanding of electromagnetism with a practical, inventive approach to engineering that made him such a pivotal figure in the history of electricity.

CHAPTER SEVENTEEN: Hertz's Waves: Confirming Maxwell's Theory and the Birth of Radio

The triumph of alternating current, championed by Nikola Tesla and George Westinghouse, had revolutionized the generation and distribution of electricity. AC power, with its ability to be transmitted efficiently over long distances, was rapidly becoming the standard, illuminating cities and powering industries across the globe. Yet, as significant as these advancements were, they represented only part of a larger story, a story that had been set in motion by the theoretical work of James Clerk Maxwell.

Maxwell's equations, published in the 1860s, had unified electricity, magnetism, and light into a single elegant framework. They had revealed that light was an electromagnetic wave, a self-propagating disturbance of electric and magnetic fields traveling through space. But Maxwell's theory went even further, predicting the existence of electromagnetic waves beyond the visible spectrum, waves that were invisible to the human eye but fundamentally the same in nature as light.

This was a bold prediction, and at the time it was made, there was no direct experimental evidence to support it. The technology needed to generate and detect these hypothetical waves simply didn't exist. It would take another two decades, and the ingenuity of a young German physicist named Heinrich Hertz, to confirm Maxwell's extraordinary insight and usher in the era of wireless communication.

Heinrich Rudolf Hertz was born in Hamburg in 1857, into a prominent and well-educated family. His father was a lawyer and later a senator, and his mother was the daughter of a physician. Hertz showed an early aptitude for science and languages, and he excelled in his studies. He initially pursued a career in engineering, but his fascination with fundamental physics soon led him to switch his focus to pure science.

Hertz studied physics at the University of Berlin under the renowned physicists Hermann von Helmholtz and Gustav Kirchhoff. Helmholtz, in particular, had a profound influence on Hertz. Helmholtz was a leading figure in the field of electromagnetism, and he recognized Hertz's exceptional talent. He encouraged Hertz to pursue research on Maxwell's theory, suggesting that it might be possible to generate and detect electromagnetic waves experimentally.

After receiving his doctorate in 1880, Hertz continued his research, first at the University of Kiel and later at the Karlsruhe Polytechnic. He was particularly interested in the phenomenon of electrical oscillations, the rapid back-and-forth flow of electric current in a circuit. He realized that these oscillations might be the key to generating the electromagnetic waves predicted by Maxwell.

Maxwell's theory suggested that accelerating electric charges would produce electromagnetic waves. Hertz reasoned that a rapidly oscillating electric current, in which the charges were constantly accelerating and decelerating, should therefore emit these waves. The challenge was to create a circuit that could produce oscillations of sufficiently high frequency to generate detectable waves.

Hertz began experimenting with various types of oscillators, including Leyden jars and induction coils. An induction coil is a device that uses electromagnetic induction to produce high-voltage pulses from a low-voltage direct current source. It consists of two coils of wire wound around a common iron core. When a direct current flowing through the primary coil is interrupted, it creates a rapidly changing magnetic field in the core, which induces a high-voltage pulse in the secondary coil.

Hertz modified the induction coil to create a spark gap oscillator. He connected the secondary coil to two metal rods separated by a small gap. When the induction coil produced a high-voltage pulse, it caused a spark to jump across the gap, creating a brief, intense burst of oscillating current.

Hertz theorized that this spark, with its rapidly oscillating charges, should emit electromagnetic waves. But how could he detect these invisible waves? He needed a receiver, a device that could somehow sense the presence of the waves and convert their energy back into a detectable electrical signal.

Hertz's solution was ingenious and surprisingly simple. He fashioned a loop of wire with a small gap, similar to the spark gap in his oscillator. He reasoned that if electromagnetic waves were present, they would induce an oscillating current in the loop, and if the frequency of the waves matched the natural frequency of the loop, a spark would jump across the gap, mirroring the spark in the oscillator.

In 1886, Hertz conducted his first successful experiments with his spark-gap transmitter and loop receiver. He placed the oscillator in one part of his laboratory and the receiver in another, several meters away. To his astonishment, when he activated the oscillator, he observed tiny sparks jumping across the gap in the receiver. This was a momentous discovery. He had generated and detected electromagnetic waves, confirming Maxwell's prediction.

Hertz's initial experiments were just the beginning. He conducted a series of meticulous investigations, exploring the properties of these newly discovered waves. He showed that they could be reflected, refracted, and diffracted, just like light waves. He also demonstrated that they could be polarized, meaning that their electric and magnetic fields oscillated in specific directions, another property shared with light.

One of Hertz's most important experiments involved measuring the speed of the waves. By creating standing waves, a phenomenon that occurs when waves are reflected back on themselves, he was able to measure the wavelength of the electromagnetic waves. Knowing the frequency of his oscillator, he could then calculate the speed of the waves using the simple formula: speed = frequency × wavelength.

Hertz's measurements showed that the speed of the electromagnetic waves was approximately equal to the speed of light, just as Maxwell had predicted. This was compelling evidence that these waves were indeed a form of electromagnetic radiation, fundamentally the same as light but with a much longer wavelength.

Hertz's experiments were a triumph of experimental physics. They provided the first definitive proof of the existence of electromagnetic waves beyond the visible spectrum, confirming Maxwell's theory and revolutionizing our understanding of the nature of light and electromagnetism.

Hertz published his findings in a series of papers between 1887 and 1889, culminating in his famous paper, "On Electromagnetic Waves in Air and Their Reflection." His work created a sensation in the scientific community, and it was quickly recognized as a landmark achievement.

Hertz's discovery had profound implications. It opened up a whole new realm of physics, the study of the electromagnetic spectrum beyond visible light. It also laid the foundation for the development of wireless communication, although Hertz himself did not fully appreciate the practical potential of his discovery.

When asked about the practical applications of his waves, Hertz famously replied, "It's of no use whatsoever... this is just an experiment that proves Maestro Maxwell was right—we just have these mysterious electromagnetic waves that we cannot see with the naked eye. But they are there."

Despite his skepticism about the practical uses of his discovery, Hertz's work sparked the imagination of other scientists and inventors. They recognized that if electromagnetic waves could be generated and detected, they could also be used to transmit information wirelessly, without the need for connecting wires.

One of the first to grasp the potential of wireless communication was the British physicist Oliver Lodge. Lodge had been working independently on electromagnetic waves, and he was deeply impressed by Hertz's experiments. In 1894, he gave a lecture at the Royal Institution in London, where he demonstrated a system for transmitting and receiving signals using Hertzian waves, as they were then called.

Lodge's system used a modified version of Hertz's spark-gap transmitter and a device called a coherer as a receiver. The coherer, invented by the French physicist Édouard Branly, was a glass tube filled with metal filings. Normally, the filings had a high resistance to electric current. However, when exposed to electromagnetic waves, the filings would "cohere," or stick together, reducing their resistance and allowing a current to flow.

Lodge's demonstration was a significant step towards practical wireless communication. He showed that signals could be transmitted over short distances using electromagnetic waves, and he even suggested that it might be possible to use this technology for telegraphy without wires.

However, it was the Italian inventor Guglielmo Marconi who would ultimately transform Hertz's discovery into a practical and commercially successful wireless communication system. Marconi, born in 1874, was a self-taught engineer with a keen interest in electricity and radio waves. He was inspired by Hertz's and Lodge's work, and he set out to develop a system that could transmit signals over long distances.

Marconi began experimenting with wireless telegraphy in 1895, at his family's estate in Italy. He improved upon the existing technology, increasing the power of the transmitter, the sensitivity of the receiver, and the height of the antennas. He also developed a method for tuning the transmitter and receiver to the same frequency, which greatly improved the reliability of the system.

Marconi's early experiments were successful, and he gradually increased the range of his transmissions. In 1896, he moved to England, where he continued to develop his system and sought financial backing for his work. He gave several public demonstrations of wireless telegraphy, attracting the attention of the British Post Office, which was interested in using the technology for ship-to-shore communication.

In 1897, Marconi established the Wireless Telegraph and Signal Company, later known as Marconi's Wireless Telegraph Company, to develop and commercialize his invention. He continued to improve his system, increasing the range and reliability of his transmissions.

One of Marconi's most significant achievements was the first transatlantic wireless transmission in 1901. He set up a powerful transmitter in Poldhu, Cornwall, England, and a receiving station in St. John's, Newfoundland, Canada. On December 12, 1901, he successfully transmitted the Morse code letter "S" across the Atlantic Ocean, a distance of over 2,000 miles.

This was a landmark event in the history of communication. It demonstrated that electromagnetic waves could travel vast distances, even across oceans, and it opened up the possibility of instantaneous global communication. Marconi's transatlantic transmission captured the public's imagination, and it marked the beginning of the radio age.

The development of wireless telegraphy, or radio as it came to be known, was a transformative moment in human history. It revolutionized communication, making it possible to transmit information across vast distances without the need for wires. It had a profound impact on shipping, enabling ships to communicate with each other and with shore stations, improving safety and navigation. It also transformed warfare, providing a new means of communication for military forces.

Radio also became a powerful medium for news and entertainment. The first radio broadcasts began in the early 20th century, and radio quickly became a popular form of mass communication. It brought news, music, and entertainment into people's homes, transforming their lives and creating a new form of shared cultural experience.

The development of radio was not solely the work of Marconi. Many other scientists and inventors contributed to its development, including Reginald Fessenden, who pioneered the transmission of voice and music over radio, and Lee de Forest, who invented the Audion, the first practical electronic amplifier, which greatly improved the sensitivity of radio receivers.

The invention of the Audion, a type of vacuum tube, was a crucial step in the development of electronics. It allowed for the amplification of weak electrical signals, making it possible to receive faint radio waves and to build more sophisticated electronic circuits. The Audion and other vacuum tubes became the fundamental building blocks of electronic devices for the first half of the 20th century, until they were eventually replaced by transistors.

Heinrich Hertz, the man who had started it all with his groundbreaking experiments on electromagnetic waves, did not live to see the full impact of his discovery. He died in 1894 at the young age of 36, from complications of an infection. However, his legacy lived on, and his name is forever associated with the discovery of electromagnetic waves and the birth of radio.

The unit of frequency, the hertz (Hz), is named in his honor. One hertz is equal to one cycle per second, a fitting tribute to the man who first measured the frequency of electromagnetic waves. Hertz's work not only confirmed Maxwell's theory but also opened up a whole new field of physics and engineering.

Hertz's experiments were a triumph of scientific inquiry, a testament to the power of human curiosity and the importance of experimental verification. His discovery of electromagnetic waves was a pivotal moment in the history of science, a turning point that transformed our understanding of the universe and paved the way for technologies that have reshaped the modern world. His legacy continues to inspire scientists and engineers today, as they explore the vast and complex electromagnetic spectrum and seek new ways to harness its power for the benefit of humanity. The story of Hertz and his waves is a reminder that fundamental scientific discoveries, even those that seem abstract and far removed from everyday life, can have profound and unexpected practical consequences, shaping the course of history in ways that their discoverers could never have imagined.

CHAPTER EIGHTEEN: The Electron's Discovery: Unveiling the Fundamental Unit of Charge

The late 19th century was a golden age for physics. The triumphs of classical mechanics, thermodynamics, and Maxwell's electromagnetic theory had created a sense of completeness, a feeling that the fundamental laws governing the physical world had been discovered. Yet, beneath this আপাত surface of understanding, subtle cracks were beginning to appear. Experiments with electricity and magnetism, particularly those involving the behavior of electricity in gases, were revealing phenomena that could not be fully explained by existing theories. These unexplained observations would ultimately lead to a revolution in physics, a revolution that began with the discovery of the electron, the fundamental unit of electric charge.

For centuries, electricity had been thought of as a continuous fluid, a substance that could flow through conductors and accumulate on surfaces. This fluidic model had been successful in explaining many electrical phenomena, such as the flow of current in circuits and the behavior of capacitors. However, it failed to account for the growing body of evidence suggesting that electricity might have a more granular, particulate nature.

One of the key areas of research that challenged the fluidic model of electricity was the study of cathode rays. Cathode rays were mysterious emanations produced in vacuum tubes when a high voltage was applied between two electrodes. These tubes, also known as Crookes tubes or Geissler tubes, were glass vessels from which most of the air had been evacuated. When a high voltage was applied across the electrodes, a luminous glow appeared in the tube, and strange rays seemed to emanate from the negative electrode, the cathode.

The nature of cathode rays was a subject of intense debate among physicists in the late 19th century. Some believed that they were a form of electromagnetic radiation, similar to light but with a shorter wavelength. Others argued that they were streams of particles, a hypothesis that seemed to be supported by the observation that cathode rays could be deflected by magnetic fields.

One of the early pioneers in the study of cathode rays was Julius Plücker, a German mathematician, and physicist. In the 1850s, Plücker conducted experiments with Geissler tubes, which were evacuated glass tubes containing various gases at low pressure. He observed that when a high voltage was applied across the electrodes, the gas inside the tube would glow with different colors depending on the type of gas. He also found that the glow could be deflected by a magnetic field, suggesting that it was associated with charged particles.

Plücker's work was continued by his student, Johann Wilhelm Hittorf, who made several important improvements to the design of vacuum tubes. Hittorf introduced a better vacuum pump, which allowed him to achieve lower pressures in the tubes. He also added a second electrode, called an anode, to collect the rays emitted by the cathode.

Hittorf's experiments revealed that cathode rays traveled in straight lines and cast sharp shadows when they encountered an obstacle. This observation supported the hypothesis that cathode rays were composed of particles, as waves would typically diffract around obstacles rather than casting sharp shadows.

Further evidence for the particle nature of cathode rays came from the work of William Crookes, an English physicist, and chemist. Crookes developed improved vacuum tubes, which he called Crookes tubes, that could achieve even lower pressures than Hittorf's tubes. He also introduced various objects into the tubes, such as Maltese crosses and paddle wheels, to study the effects of cathode rays.

Crookes' experiments showed that cathode rays could exert mechanical force, as evidenced by the rotation of the paddle wheel when placed in their path. He also found that the rays could be deflected by both electric and magnetic fields, confirming that they carried an electric charge. Crookes believed that cathode rays were composed of negatively charged "radiant matter," a fourth state of matter beyond solid, liquid, and gas.

The debate over the nature of cathode rays continued for several decades. While the evidence for their particle nature was mounting, some physicists still clung to the idea that they were a form of electromagnetic radiation. The key to resolving this debate would be to measure the charge-to-mass ratio of the particles, a quantity that could potentially distinguish between different types of particles and shed light on their fundamental nature.

This is where J.J. Thomson, a British physicist, entered the scene. Thomson, who was born in 1856 near Manchester, England, was a brilliant scientist who would go on to become one of the most influential physicists of his time. He studied at Owens College (now the University of Manchester) and later at Trinity College, Cambridge, where he excelled in mathematics and physics.

In 1884, at the young age of 28, Thomson was appointed Cavendish Professor of Experimental Physics at Cambridge University, a prestigious position that he would hold for the next 34 years. As head of the Cavendish Laboratory, Thomson transformed it into a world-leading center for research in physics, attracting talented students and researchers from around the globe.

Thomson was particularly interested in the study of electricity and magnetism, and he was intrigued by the ongoing debate over the nature of cathode rays. He set out to design experiments that could definitively determine whether these rays were particles or waves and, if they were particles, to measure their properties.

Thomson's experimental setup was a masterpiece of ingenuity and precision. He used a modified Crookes tube with a fluorescent screen at one end. The cathode rays, emitted from the cathode, were accelerated towards the anode, which had a small hole in its center. Some of the rays passed through the hole, forming a narrow beam that continued towards the fluorescent screen, creating a bright spot where it struck.

To determine the nature of the cathode rays, Thomson subjected the beam to both electric and magnetic fields. He placed two parallel metal plates on either side of the beam to create an electric field, and he used a pair of coils to generate a magnetic field perpendicular to the beam's path.

Thomson's reasoning was as follows: if the cathode rays were particles, they should be deflected by both electric and magnetic fields. The direction and magnitude of the deflection would depend on the charge and mass of the particles, as well as the strength of the fields. By carefully measuring the deflection of the beam under different field conditions, he hoped to determine the charge-to-mass ratio (e/m) of the particles.

Thomson's first experiments confirmed that cathode rays were indeed deflected by both electric and magnetic fields, providing further evidence for their particle nature. He then set out to measure the charge-to-mass ratio. He adjusted the strengths of the electric and magnetic fields so that their effects on the beam canceled each other out, resulting in no net deflection. By equating the electric and magnetic forces on the particles, he could derive an expression for their velocity.

Next, Thomson measured the deflection of the beam caused by the magnetic field alone. Knowing the velocity of the particles and the strength of the magnetic field, he could calculate the radius of curvature of their path. This, in turn, allowed him to determine the charge-to-mass ratio using the formula for the magnetic force on a moving charge in a magnetic field.

Thomson's measurements of the charge-to-mass ratio of cathode rays yielded a startling result. The value he obtained was much larger than that of any known ion, about 1,800 times greater than that of the hydrogen ion, the lightest known ion at the time. This meant that either the charge of the cathode ray particles was much larger than that of ions, or their mass was much smaller.

Thomson conducted further experiments to refine his measurements and to rule out the possibility that the high charge-to-mass ratio was due to an unusually large charge. He varied the material of the cathode and the type of gas in the tube, but the charge-to-mass ratio remained constant. This suggested that the cathode ray particles were a fundamental constituent of all matter, not specific to a particular element.

In 1897, Thomson announced his groundbreaking conclusion: cathode rays were not a form of electromagnetic radiation, but rather streams of negatively charged particles that were much smaller and lighter than atoms. He called these particles "corpuscles," but they soon became known as "electrons," a term that had been previously suggested by the Irish physicist George Johnstone Stoney for the fundamental unit of electric charge.

Thomson's discovery of the electron was a watershed moment in the history of physics. It was the first subatomic particle to be identified, and it revolutionized our understanding of the nature of matter and electricity. It showed that the atom, which had been considered the smallest indivisible unit of matter, was actually composed of even smaller particles.

Thomson proposed a model of the atom, often called the "plum pudding" model, in which negatively charged electrons were embedded in a sphere of positive charge, like plums in a pudding. This model was a significant departure from the earlier idea of the atom as a solid, indivisible sphere.

The discovery of the electron also provided a new understanding of the nature of electric current. Instead of being a continuous fluid, electricity was now understood to be the flow of discrete, negatively charged particles. This explained why electric charge was quantized, meaning that it always appeared in multiples of a fundamental unit, the charge of the electron.

Thomson's work had far-reaching implications for physics and chemistry. It paved the way for the development of atomic physics and the study of the structure and behavior of atoms. It also led to a deeper understanding of chemical bonding and the properties of materials.

The discovery of the electron also had practical consequences. It laid the foundation for the field of electronics, which would revolutionize technology in the 20th century. The ability to control and manipulate the flow of electrons in vacuum tubes and, later, in solid-state devices like transistors, would lead to the development of radio, television, computers, and countless other electronic devices that we take for granted today.

Thomson's work earned him widespread recognition and numerous honors. He was awarded the Nobel Prize in Physics in 1906 "in recognition of the great merits of his theoretical and experimental investigations on the conduction of electricity by gases." He was also knighted in 1908, becoming Sir J.J. Thomson.

Thomson's legacy extends beyond his own discoveries. As head of the Cavendish Laboratory, he mentored a generation of physicists who would go on to make their own groundbreaking contributions to science. Seven of his research assistants, as well as his own son, George Paget Thomson, would eventually win Nobel Prizes, a testament to his influence as a teacher and a researcher.

The discovery of the electron was a pivotal moment in the history of science. It marked the beginning of a new era in physics, an era in which the atom and its constituents would become the focus of intense investigation. It opened up a whole new world of subatomic particles and forces, a world that would challenge and ultimately transform our understanding of the fundamental nature of reality.

Thomson's work was a triumph of experimental physics, a testament to the power of careful observation, precise measurement, and ingenious experimental design. It was also a triumph of scientific reasoning, the ability to draw profound conclusions from subtle clues and to construct a coherent and revolutionary theory based on experimental evidence.

The electron, once a hypothetical entity, had become a tangible reality, a fundamental building block of the universe. Its discovery would not only reshape our understanding of matter and electricity but also pave the way for technological innovations that would transform society in ways that Thomson himself could scarcely have imagined. The tiny particle that he had coaxed from the depths of the atom would illuminate the path towards a new era of scientific and technological progress, an era that continues to unfold to this day. It had brought science to the edge of a new frontier, the exploration of the subatomic world, and had provided the first glimpse of the rich and complex structure that lay hidden beneath the আপাত surface of matter. The electron, the fundamental unit of electric charge, had been revealed, and the world of physics would never be the same again.

CHAPTER NINETEEN: The Rise of Electronics: Vacuum Tubes and the Amplification of Signals

The discovery of the electron by J.J. Thomson in 1897 had profound implications for the understanding of electricity and the structure of matter. It revealed that electricity was not a continuous fluid, but rather a flow of discrete, negatively charged particles. This discovery opened up a new realm of scientific inquiry, the subatomic world, and paved the way for a new era of technology: the age of electronics.

The first major step into this new era came with the invention of the vacuum tube, a device that could control and manipulate the flow of electrons in a vacuum. Vacuum tubes, also known as electron tubes or valves, would become the fundamental building blocks of electronic devices for the first half of the 20th century. They would enable the development of radio, television, radar, computers, and a host of other technologies that would transform communication, entertainment, industry, and warfare.

The story of the vacuum tube begins with the "Edison effect," an observation made by Thomas Edison in 1883 while he was working on improving his incandescent light bulb. Edison noticed that a dark deposit would form on the inside of the glass bulb, opposite the filament. To investigate this phenomenon, he inserted a metal plate inside the bulb and connected it to a galvanometer, an instrument for detecting electric current.

Edison observed that when the plate was connected to the positive terminal of the filament's power supply, a current flowed through the galvanometer. However, when the plate was connected to the negative terminal, no current flowed. This was a puzzling observation, as there was no physical connection between the filament and the plate, only a vacuum.

Edison, who was primarily interested in practical applications rather than fundamental research, did not fully understand the significance of his observation. He patented the device, calling it an "electrical indicator," but he saw no immediate use for it and moved on to other projects.

The Edison effect, as it came to be known, was essentially the first observation of thermionic emission, the flow of electrons from a heated metal surface. The hot filament in Edison's bulb was emitting electrons, which were then attracted to the positively charged plate, creating a current. However, when the plate was negatively charged, it repelled the electrons, and no current flowed.

Edison's observation remained largely a scientific curiosity for several years. However, other researchers began to investigate the phenomenon, and they gradually came to understand its underlying principles. One of these researchers was John Ambrose Fleming, a British electrical engineer and physicist who had worked as a consultant for Edison's company in England.

Fleming, who was also a professor at University College London, was interested in the development of wireless telegraphy, or radio. He recognized that the Edison effect could potentially be used to create a more sensitive detector for radio waves than the coherers that were commonly used at the time.

In 1904, Fleming developed the first practical application of the Edison effect: the thermionic diode, also known as the Fleming valve. The Fleming valve was a vacuum tube containing two electrodes: a heated filament, or cathode, and a metal plate, or anode. The cathode was coated with a material that readily emitted electrons when heated, such as barium oxide or strontium oxide.

When the cathode was heated and a positive voltage was applied to the anode, electrons emitted from the cathode were attracted to the anode, creating a current flow through the tube. However, when the voltage on the anode was negative, it repelled the electrons, and no current flowed. This one-way flow of current made the diode useful as a rectifier, a device that converts alternating current (AC) to direct current (DC).

Fleming's diode was a significant improvement over the coherer as a detector for radio waves. It was more sensitive, more reliable, and could respond to higher frequencies. It quickly became a standard component in early radio receivers, allowing for the detection of weaker signals and improving the overall performance of the receivers.

The invention of the diode was a major step forward in the development of electronics, but it was only the beginning. The next crucial breakthrough came with the addition of a third electrode to the vacuum tube, creating the triode. This seemingly simple modification would transform the vacuum tube from a simple rectifier into a powerful amplifier, capable of boosting weak electrical signals.

The inventor of the triode was Lee de Forest, an American inventor and radio pioneer. De Forest, who was born in 1873 in Council Bluffs, Iowa, was a prolific inventor with a keen interest in wireless communication. He received a doctorate in physics from Yale University in 1899, with a dissertation on the reflection of Hertzian waves.

De Forest recognized the limitations of the diode as a radio detector and sought to improve upon its design. He experimented with adding a third electrode to the tube, placing a grid of fine wire mesh between the cathode and the anode. He called his invention the "Audion."

In 1906, De Forest discovered that by applying a small voltage to the grid, he could control the flow of electrons between the cathode and the anode. A small change in the grid voltage produced a large change in the current flowing through the tube. This meant that the Audion could act as an amplifier, taking a weak electrical signal applied to the grid and producing a much stronger version of the signal in the anode circuit.

The ability to amplify electrical signals was a game-changer for electronics. It meant that weak radio signals, which were previously difficult to detect, could now be amplified to a level where they could be easily heard. It also opened up the possibility of long-distance telephony, as telephone signals, which weaken over long distances, could now be boosted along the way.

De Forest's Audion was initially met with skepticism by some in the scientific and engineering community. Its operation was not fully understood, and its performance was often erratic. However, De Forest continued to refine his design, and he demonstrated its capabilities in a series of public demonstrations.

One of De Forest's most famous demonstrations was the broadcast of the voice of the opera singer Enrico Caruso from the Metropolitan Opera House in New York City in 1910. This was one of the first public demonstrations of live radio broadcasting, and it captured the public's imagination. It showed that radio was not just for transmitting Morse code, but that it could also be used to transmit music and the human voice.

Despite the success of his demonstrations, De Forest struggled to commercialize his invention. He faced legal battles over patent rights, and he lacked the business acumen to compete with larger, more established companies. However, the importance of his invention was eventually recognized, and the Audion, or triode as it came to be known, became a fundamental component of radio receivers and other electronic devices.

The development of the triode was a pivotal moment in the history of electronics. It was the first electronic device capable of amplification, and it opened up a whole new world of possibilities for manipulating and controlling electrical signals. It made possible the development of more sensitive and selective radio receivers, long-distance telephone systems, public address systems, and many other electronic devices.

The triode was not just important for its practical applications; it also played a crucial role in the development of the science of electronics. The ability to amplify weak signals allowed researchers to study electrical phenomena that were previously undetectable. It also led to a deeper understanding of the behavior of electrons in a vacuum and the interaction between electrons and electromagnetic fields.

The early triodes were relatively crude devices, with limited amplification capabilities and a short lifespan. However, engineers and scientists continued to refine their design, improving their performance and reliability. They experimented with different cathode materials, grid designs, and tube geometries. They also developed techniques for creating higher vacuums in the tubes, which improved their performance and longevity.

One important improvement was the development of the indirectly heated cathode. In early triodes, the filament served as both the source of heat and the source of electrons. This meant that the cathode had to be heated with a separate power supply, which added complexity and bulk to the circuit.

The indirectly heated cathode, which was developed in the 1920s, used a separate heater element to heat a cylindrical cathode coated with an electron-emitting material. This allowed the cathode to be heated with alternating current, simplifying the power supply requirements and making the tubes more suitable for use in a wider range of applications.

Another significant development was the invention of the tetrode and the pentode, which added additional grids to the triode. The tetrode, which had a screen grid between the control grid and the anode, reduced the capacitance between the control grid and the anode, improving the tube's high-frequency performance. The pentode, which added a suppressor grid between the screen grid and the anode, further improved the tube's performance by reducing the effects of secondary emission, a phenomenon in which electrons striking the anode cause the emission of additional electrons.

These improved vacuum tubes became essential components in a wide range of electronic devices. They were used in radio receivers and transmitters, audio amplifiers, telephone repeaters, and early television systems. They also found applications in industrial control systems, scientific instruments, and medical equipment.

The development of vacuum tube technology also spurred the growth of the electronics industry. Companies like General Electric, Westinghouse, and RCA invested heavily in vacuum tube research and manufacturing. They established large-scale production facilities to meet the growing demand for tubes, and they developed sophisticated manufacturing techniques to ensure the quality and reliability of their products.

The vacuum tube era reached its peak during World War II. Vacuum tubes were used extensively in military equipment, including radar, communication systems, and early computers. The war effort drove rapid advancements in vacuum tube technology, as engineers sought to improve the performance and reliability of tubes for demanding military applications.

One of the most significant wartime developments was the miniaturization of vacuum tubes. Smaller tubes were needed for portable radios, aircraft communication systems, and other compact electronic devices. Engineers developed techniques for manufacturing miniature tubes with extremely small dimensions, allowing for the construction of more compact and lightweight electronic equipment.

The war also spurred the development of specialized tubes for specific applications. For example, the magnetron, a type of vacuum tube that generates high-power microwaves, was developed for use in radar systems. The magnetron was a crucial component of Allied radar systems during the war, and it played a significant role in the development of microwave technology after the war.

The development of the electronic computer during and after World War II was another major driver of vacuum tube technology. Early computers, such as the ENIAC (Electronic Numerical Integrator and Computer), used thousands of vacuum tubes as switching and amplifying elements. The ENIAC, which was completed in 1946, contained over 17,000 vacuum tubes and occupied an entire room.

These early computers were incredibly powerful for their time, but they were also bulky, expensive, and consumed a lot of power. The heat generated by the thousands of vacuum tubes required elaborate cooling systems, and the tubes themselves were prone to failure, requiring frequent replacement.

Despite their limitations, these early vacuum tube computers demonstrated the potential of electronic computation, and they spurred further research into computer technology. The development of smaller, more reliable, and more energy-efficient vacuum tubes played a crucial role in the evolution of computers in the postwar era.

The vacuum tube era, however, was not to last forever. The seeds of its demise were sown in 1947 with the invention of the transistor, a solid-state device that could perform the same functions as a vacuum tube but was much smaller, more reliable, and consumed less power. The transistor, which will be discussed in more detail in the next chapter, would eventually replace the vacuum tube as the fundamental building block of electronic devices, ushering in the era of solid-state electronics.

However, the vacuum tube's legacy extends far beyond its relatively brief reign as the dominant electronic technology. The vacuum tube era was a period of rapid innovation and discovery, a time when the foundations of modern electronics were laid. The development of the vacuum tube not only enabled the creation of new technologies like radio, television, and computers but also fostered a deeper understanding of the behavior of electrons and the nature of electricity itself.

The vacuum tube era also saw the growth of a new industry, the electronics industry, which would become one of the most important and dynamic sectors of the global economy. The companies that pioneered vacuum tube technology, such as General Electric, Westinghouse, and RCA, became major players in the electronics industry, and their research laboratories made significant contributions to the advancement of science and technology.

Moreover, the vacuum tube era trained a generation of engineers and scientists in the principles of electronics. The knowledge and skills acquired during this period would prove invaluable in the development of solid-state electronics and the subsequent explosion of digital technology. Many of the pioneers of the transistor era, including William Shockley, one of the inventors of the transistor, had their roots in the vacuum tube era.

The story of the vacuum tube is a testament to the power of human ingenuity and the transformative potential of scientific discovery. From Edison's initial observation of thermionic emission to De Forest's invention of the triode and the subsequent development of a vast array of specialized tubes, the evolution of the vacuum tube was a remarkable journey of innovation. It was a journey driven by both scientific curiosity and the desire to create new and useful technologies.

The vacuum tube may now be a relic of the past, largely replaced by smaller, faster, and more efficient solid-state devices. However, its impact on the world is undeniable. It was the vacuum tube that amplified the first faint radio signals, bringing music and voices into people's homes. It was the vacuum tube that powered the first electronic computers, laying the foundation for the digital age. And it was the vacuum tube that trained a generation of engineers and scientists, equipping them with the knowledge and skills to continue pushing the boundaries of electronics. The humble vacuum tube, a seemingly simple device consisting of a few electrodes sealed in a glass envelope, was a giant leap forward for technology, a leap that would forever change the course of human history. It was a crucial stepping stone on the path to our modern, interconnected world, a world where electronics play an integral role in nearly every aspect of our lives.

CHAPTER TWENTY: The Transistor Revolution: Miniaturization and the Digital Age

The vacuum tube, despite its revolutionary impact on the development of electronics, had inherent limitations. Tubes were bulky, fragile, consumed significant amounts of power, and generated a lot of heat. They also had a limited lifespan, burning out like light bulbs and requiring regular replacement. These limitations became increasingly problematic as electronic devices grew more complex, particularly in the burgeoning field of computing. The ENIAC, one of the first electronic general-purpose computers, contained over 17,000 vacuum tubes, weighed 30 tons, occupied 1,800 square feet of floor space, and consumed 150 kilowatts of power. It was clear that a new technology was needed to replace the vacuum tube if electronics were to continue advancing.

That new technology arrived in 1947 with the invention of the transistor, a small, solid-state device that could perform the same functions as a vacuum tube—amplification and switching—but with far greater efficiency and reliability. The transistor would not only replace the vacuum tube, but it would also revolutionize electronics, enabling the miniaturization of circuits and the development of the digital age.

The story of the transistor begins at Bell Telephone Laboratories, the research arm of the American Telephone and Telegraph Company (AT&T). Bell Labs, as it was commonly known, was one of the premier industrial research institutions in the world, with a long history of innovation in communications technology. In the 1930s, Bell Labs' director of research, Mervin Kelly, recognized that the growing complexity of the telephone network would eventually require electronic switching systems to replace the existing electromechanical relays. He also understood that the vacuum tubes of the time were not well-suited for this task due to their limitations.

Kelly believed that a new type of electronic switch, based on solid-state materials rather than vacuum tubes, was needed. He envisioned a device that could control the flow of electricity in a solid material, similar to the way a valve controls the flow of water in a pipe. This vision led him to establish a research group dedicated to studying the properties of semiconductors, materials that have electrical conductivity between that of a conductor, like copper, and an insulator, like glass.

The semiconductor research group at Bell Labs was led by physicist William Shockley. Shockley, who had earned his doctorate from MIT in 1936, was a brilliant and ambitious scientist with a deep understanding of solid-state physics. He assembled a team of talented researchers, including John Bardeen, an expert in the theory of electrons in solids, and Walter Brattain, a skilled experimentalist.

The team's initial focus was on understanding the behavior of electrons in semiconductors, particularly silicon and germanium. These elements, which belong to the same group in the periodic table as carbon, have unique electrical properties that make them suitable for use in electronic devices.

Semiconductors are characterized by their ability to conduct electricity under certain conditions but not others. Unlike metals, which have a large number of free electrons that can move easily, semiconductors have a relatively small number of free electrons. However, the conductivity of a semiconductor can be dramatically altered by adding small amounts of impurities, a process known as doping.

Doping involves introducing atoms of a different element into the semiconductor crystal lattice. These impurity atoms can either donate extra electrons to the material, creating an n-type semiconductor, or they can create "holes,' which are effectively the absence of an electron and act as positive charge carriers, creating a p-type semiconductor.

The behavior of electrons and holes at the junction between n-type and p-type semiconductors is the key to the operation of many solid-state devices, including the transistor. When an n-type and a p-type semiconductor are brought together, they form a p-n junction. At this junction, electrons from the n-type material diffuse into the p-type material, and holes from the p-type material diffuse into the n-type material. This creates a depletion region, a zone near the junction that is depleted of free charge carriers.

The depletion region acts as a barrier to the flow of current. However, the width of this barrier, and therefore the conductivity of the junction, can be controlled by applying an external voltage. This ability to control the flow of current through a p-n junction is the fundamental principle behind the operation of diodes and transistors.

Shockley's team initially focused on developing a type of transistor called a field-effect transistor (FET). The basic idea behind a FET is to use an electric field to control the conductivity of a semiconductor channel. A voltage applied to a "gate" electrode creates an electric field that modulates the number of charge carriers in the channel, thereby controlling the current flow between two other electrodes, the "source" and the "drain."

Despite their efforts, Shockley's team struggled to create a working FET. They encountered numerous technical challenges, including the difficulty of creating a sufficiently strong electric field within the semiconductor material and the presence of surface states, defects on the surface of the semiconductor that trapped charge carriers and interfered with the operation of the device.

While working on the FET, Bardeen and Brattain began experimenting with a different type of transistor, one that relied on the properties of p-n junctions rather than electric fields. They fabricated a device consisting of a small block of germanium with two closely spaced gold contacts on one surface and a larger contact on the opposite surface.

On December 16, 1947, Bardeen and Brattain made a breakthrough. They discovered that by applying a small current to one of the gold contacts, they could control a much larger current flowing between the other gold contact and the base. This was the first demonstration of transistor action, the ability of a small electrical signal to control a larger one. They had invented the point-contact transistor.

The point-contact transistor was a crude device, and its operation was not fully understood at the time. However, it was a proof of concept, a demonstration that a solid-state device could amplify electrical signals. This was a watershed moment in the history of electronics, the birth of the solid-state era.

News of the invention spread quickly within Bell Labs, and Shockley, although initially disappointed that he had not been directly involved in the discovery, recognized its significance. He immediately set out to develop a more robust and practical transistor design.

Within a few weeks, Shockley conceived of the junction transistor, a device that would become the dominant type of transistor for decades to come. Unlike the point-contact transistor, which relied on surface effects, the junction transistor was based on the properties of p-n junctions within the bulk of the semiconductor material.

Shockley's junction transistor consisted of a sandwich of three layers of doped semiconductor material, either in an n-p-n or a p-n-p configuration. The outer layers served as the emitter and collector, while the middle layer, which was much thinner, served as the base.

The operation of the junction transistor is based on the ability of a small current injected into the base to control a much larger current flowing between the emitter and the collector. When a small forward bias is applied to the base-emitter junction, it allows electrons (in an n-p-n transistor) or holes (in a p-n-p transistor) to flow from the emitter into the base. Because the base is very thin, most of these charge carriers diffuse across the base and are swept into the collector by the electric field at the base-collector junction.

The key to the transistor's amplifying action is that a small change in the base current produces a much larger change in the collector current. This is because the collector current is proportional to the number of charge carriers that reach the collector, which in turn is controlled by the base current.

Shockley's junction transistor was a major improvement over the point-contact transistor. It was more stable, more reliable, and its operation could be more easily explained using the principles of semiconductor physics. It also had the potential for mass production, as its structure was more amenable to fabrication using techniques like diffusion and lithography.

Bell Labs announced the invention of the transistor in June 1948, and the news created a stir in the scientific and engineering community. However, it took several years for the transistor to move from a laboratory curiosity to a practical technology. The early transistors were difficult to manufacture, and their performance was often inconsistent.

One of the key challenges was the purification of the semiconductor materials. Transistors required extremely pure silicon or germanium, with impurity levels of less than one part per billion. This level of purity was difficult to achieve with the existing refining techniques.

Another challenge was the development of reliable fabrication processes. The first transistors were made by hand, one at a time, a process that was both time-consuming and expensive. To make transistors commercially viable, new manufacturing techniques were needed that could produce large quantities of devices with consistent characteristics.

Despite these challenges, progress was rapid. In the early 1950s, several companies, including Texas Instruments, began producing junction transistors commercially. The first transistor radios appeared on the market in 1954, marking the beginning of the consumer electronics revolution.

The advantages of transistors over vacuum tubes were immediately apparent. Transistors were much smaller, consumed far less power, and generated less heat. They were also more rugged and had a much longer lifespan. These advantages made them ideal for use in portable devices like radios and hearing aids, as well as in more complex systems like computers.

The development of the transistor also spurred research into other semiconductor devices. In 1956, Shockley, Bardeen, and Brattain were awarded the Nobel Prize in Physics for their invention of the transistor, recognizing the profound impact of their work on science and technology.

The transistor's impact on electronics was transformative. It enabled the miniaturization of electronic circuits, leading to the development of smaller, more powerful, and more affordable electronic devices. It also paved the way for the digital revolution, as transistors could be used not only as amplifiers but also as switches, the basic building blocks of digital logic circuits.

One of the key developments that accelerated the adoption of transistors was the invention of the integrated circuit (IC) in the late 1950s. The integrated circuit, also known as a microchip, is a single semiconductor crystal, usually silicon, that contains multiple interconnected transistors and other electronic components, such as resistors and capacitors.

The first integrated circuits were developed independently by Jack Kilby of Texas Instruments and Robert Noyce of Fairchild Semiconductor in 1958-1959. Kilby's IC was a simple device containing a single transistor and a few other components on a germanium substrate. Noyce's IC was more complex, with multiple transistors and other components interconnected on a silicon substrate using a planar process.

The planar process, which was developed by Jean Hoerni at Fairchild, was a key innovation that enabled the mass production of integrated circuits. It involved creating the different layers of the IC on the surface of a flat silicon wafer using techniques like photolithography, etching, and diffusion. This allowed for the fabrication of entire circuits on a single chip, greatly reducing the size and cost of electronic devices.

The invention of the integrated circuit marked the beginning of the microelectronics era. It enabled the creation of increasingly complex electronic circuits on a single chip, leading to an exponential increase in the number of transistors that could be integrated into a single device. This trend, known as Moore's Law, has driven the rapid advancement of electronics for over half a century.

Moore's Law, which was first articulated by Gordon Moore, co-founder of Intel, in 1965, states that the number of transistors on an integrated circuit doubles approximately every two years. This observation, which has held true for decades, has been a powerful predictor of the pace of technological progress in the semiconductor industry.

The miniaturization made possible by the integrated circuit has had a profound impact on society. It has enabled the development of personal computers, smartphones, the internet, and countless other technologies that have transformed the way we live, work, and communicate. It has also driven down the cost of electronics, making these technologies accessible to billions of people around the world.

The transistor revolution, sparked by the invention of the transistor at Bell Labs in 1947, has been one of the most significant technological transformations in human history. It has replaced the bulky and inefficient vacuum tube with a small, reliable, and energy-efficient solid-state device, enabling the miniaturization of electronics and the development of the digital age. From its humble beginnings as a laboratory curiosity, the transistor has become the fundamental building block of modern electronics, powering everything from pocket calculators to supercomputers. Its impact on society has been profound and far-reaching, transforming industries, economies, and the very fabric of our daily lives. The transistor, a tiny switch made of silicon, has truly changed the world, ushering in an era of unprecedented technological progress and shaping the course of human history for generations to come. Its legacy is not merely in the devices it has enabled, but in the way it has empowered us to connect, communicate, and compute in ways that were unimaginable just a few decades ago. It is also very important to be aware that the transistor was not invented 'out of the blue' but was the result of a long chain of scientific discoveries which preceded it: the work of Faraday, Maxwell, and Hertz being particularly important.

CHAPTER TWENTY-ONE: Integrated Circuits: The Building Blocks of Modern Electronics

The invention of the transistor in 1947 revolutionized electronics, replacing bulky vacuum tubes with small, efficient, and reliable solid-state devices. However, the true potential of the transistor was not fully realized until the development of the integrated circuit (IC), also known as the microchip. The IC, a tiny sliver of semiconductor material containing thousands, millions, or even billions of interconnected transistors and other electronic components, would become the fundamental building block of modern electronics, enabling the creation of increasingly complex and powerful devices, from computers and smartphones to medical equipment and spacecraft.

The concept of the integrated circuit was born out of the "tyranny of numbers," a term coined by Jack Kilby, an engineer at Texas Instruments. In the late 1950s, electronic circuits were built by wiring together individual transistors, resistors, capacitors, and other components on a circuit board. As circuits became more complex, the number of components and interconnections grew rapidly, leading to several problems.

First, the sheer number of components and the complexity of the wiring made circuits difficult and expensive to manufacture. Each component had to be individually tested and soldered onto the circuit board, a time-consuming and labor-intensive process. Second, the large number of interconnections increased the likelihood of errors and failures. A single faulty connection could render the entire circuit inoperable. Third, the size and weight of the circuits were becoming a limiting factor, especially for applications like aerospace and portable electronics.

Kilby, who had joined Texas Instruments in 1958, recognized that the solution to the tyranny of numbers lay in miniaturization. He envisioned a circuit in which all the components, both active (like transistors) and passive (like resistors and capacitors), were fabricated on a single piece of semiconductor material, eliminating the need for separate components and manual wiring.

In the summer of 1958, while most of his colleagues were on vacation, Kilby, a new employee who wasn't entitled to vacation, began working on his idea. He realized that not only could transistors be made from a semiconductor material, such as germanium, but that it should be possible to fashion resistors and capacitors from the same material by carefully controlling the doping and geometry of different regions.

Kilby's first integrated circuit, which he built in September 1958, was a simple phase-shift oscillator, a circuit that produces an oscillating electrical signal. It consisted of a single transistor, several resistors, and a capacitor, all fabricated on a tiny sliver of germanium, about half an inch long and thinner than a toothpick. The components were isolated from each other by p-n junctions and interconnected using tiny gold wires bonded to the surface of the chip.

Kilby's device was crude, but it worked. It demonstrated that it was possible to create a complete electronic circuit on a single piece of semiconductor material. This was a watershed moment in the history of electronics, the birth of the integrated circuit.

Around the same time, Robert Noyce, a physicist and co-founder of Fairchild Semiconductor, was independently working on a similar idea. Noyce, who had been one of the first employees at Shockley Semiconductor Laboratory before leaving to co-found Fairchild, had also recognized the limitations of discrete components and the need for miniaturization.

Noyce's approach to the integrated circuit was different from Kilby's. Instead of using individual wires to connect the components, Noyce proposed using a technique called the planar process, which had been developed by his colleague Jean Hoerni at Fairchild.

The planar process involved creating the different layers of the integrated circuit on the surface of a flat silicon wafer using techniques like photolithography, etching, and diffusion. Photolithography, a process borrowed from the printing industry, allowed for the precise patterning of the wafer surface using light-sensitive materials. Etching was used to remove unwanted material, and diffusion was used to introduce dopants into specific regions of the wafer.

Noyce realized that the planar process could be used to create not only transistors but also other components like resistors and capacitors, all on the same silicon wafer. He also proposed using metal films, deposited on the surface of the wafer and patterned using photolithography, to interconnect the components, eliminating the need for individual wires.

Noyce's first integrated circuit, which he built in 1959, was more sophisticated than Kilby's. It was made on a silicon wafer using the planar process and contained four transistors interconnected by metal films. Noyce's design was more practical for mass production, as it did not require the manual bonding of wires and was more amenable to automation.

The invention of the integrated circuit by Kilby and Noyce sparked a revolution in electronics. It solved the tyranny of numbers by allowing for the creation of complex circuits on a single chip, dramatically reducing the size, weight, and cost of electronic devices. It also improved the reliability of circuits by eliminating the need for numerous external connections.

The first commercial integrated circuits were introduced in the early 1960s. These early ICs were relatively simple, containing only a few transistors and other components. They were used in niche applications, such as hearing aids and military equipment, where their small size and low power consumption were particularly advantageous.

However, the potential of the integrated circuit was quickly recognized, and companies like Texas Instruments and Fairchild

Semiconductor invested heavily in developing more advanced ICs. The number of components that could be integrated on a single chip, known as the integration density, began to increase rapidly.

In 1965, Gordon Moore, who had co-founded Fairchild with Noyce and later co-founded Intel, made a remarkable observation that would become known as Moore's Law. Moore noticed that the number of transistors on integrated circuits had been doubling approximately every year (later revised to every two years), and he predicted that this trend would continue for at least the next decade.

Moore's Law was not a physical law but rather an observation about the pace of technological progress in the semiconductor industry. It was driven by a combination of factors, including improvements in manufacturing technology, increasing wafer sizes, and advances in circuit design.

Moore's prediction turned out to be remarkably accurate. For more than half a century, the semiconductor industry has been able to double the number of transistors on a chip roughly every two years, leading to an exponential increase in the complexity and performance of integrated circuits.

The rapid progress in integrated circuit technology was driven by a virtuous cycle. As the number of transistors on a chip increased, the cost per transistor decreased, making it economically feasible to use more transistors in each new generation of devices. This, in turn, drove demand for more advanced ICs, fueling further investment in research and development.

One of the key factors driving the increase in integration density was the development of new and improved manufacturing techniques. Photolithography, the process used to pattern the wafer surface, was continually refined, allowing for the creation of smaller and smaller features. The wavelength of light used in photolithography was reduced, from visible light to ultraviolet light and eventually to extreme ultraviolet light, enabling the fabrication of features smaller than the wavelength of visible light.

Another important development was the introduction of new materials. Silicon remained the dominant semiconductor material, but other materials, such as silicon dioxide for insulation and aluminum for interconnections, were also used. New techniques were developed for depositing and etching these materials with ever-increasing precision.

The size of the silicon wafers used to fabricate ICs also increased over time, from 1 inch in diameter in the early 1960s to 12 inches (300 mm) or more today. Larger wafers allowed for more chips to be produced at once, reducing the cost per chip.

As the number of transistors on a chip increased, so did the complexity of the circuits. Early ICs contained simple logic gates, the basic building blocks of digital circuits. However, by the 1970s, it became possible to integrate entire systems on a single chip, including microprocessors, memory, and input/output circuits.

The development of the microprocessor, a central processing unit (CPU) on a single chip, was a major milestone in the history of integrated circuits. The first commercial microprocessor, the Intel 4004, was introduced in 1971. It contained 2,300 transistors and could perform about 60,000 operations per second.

The microprocessor revolutionized computing, making it possible to build smaller, cheaper, and more powerful computers. It also enabled the development of personal computers, which brought computing power to individuals and small businesses.

The integration density of microprocessors and other ICs continued to increase at an astonishing rate, following Moore's Law. By the 1980s, microprocessors contained hundreds of thousands of transistors, and by the 1990s, they contained millions. Today, the most advanced microprocessors contain billions of transistors and can perform trillions of operations per second.

The increasing complexity of integrated circuits required new design tools and methodologies. In the early days of ICs, circuits were designed by hand, with engineers drawing the layout of each layer on large sheets of paper. However, as the number of transistors increased, this manual approach became impractical.

The development of computer-aided design (CAD) tools in the 1970s and 1980s revolutionized IC design. CAD tools allowed engineers to design circuits using graphical interfaces, and they automated many of the tedious and error-prone tasks involved in layout design. They also enabled the simulation and verification of circuit designs before fabrication, reducing the risk of costly errors.

Another important development was the creation of design libraries, collections of pre-designed and pre-verified circuit blocks that could be reused in different designs. This modular approach to design greatly improved productivity and reduced design time.

The manufacturing of integrated circuits, also known as semiconductor fabrication, is one of the most complex and demanding industrial processes ever developed. It requires enormous capital investment, highly specialized equipment, and extremely clean manufacturing environments.

A typical semiconductor fabrication plant, or fab, costs billions of dollars to build and equip. It contains hundreds of specialized machines, each designed to perform a specific step in the fabrication process. These machines operate in cleanrooms, highly controlled environments where the air is filtered to remove even the tiniest particles of dust, which could contaminate the wafers and ruin the circuits.

The fabrication process begins with a bare silicon wafer, a thin, circular slice of extremely pure silicon. The wafer is typically 8 or 12 inches in diameter and less than a millimeter thick. The surface of the wafer is polished to a mirror-like finish, and it is then subjected to a series of processing steps that create the different layers of the integrated circuit.

The first step is usually the creation of a layer of silicon dioxide, which acts as an insulator. This is done by heating the wafer in an oxygen-rich atmosphere, causing a thin layer of oxide to grow on the surface.

Next, a layer of photosensitive material called photoresist is applied to the wafer. The photoresist is then exposed to light through a mask, a glass plate containing the pattern of the circuit layer being fabricated. The exposed areas of the photoresist undergo a chemical change, making them either soluble or insoluble in a developer solution.

The wafer is then immersed in the developer, which removes either the exposed or unexposed areas of the photoresist, depending on whether a positive or negative photoresist is used. This leaves behind a pattern of photoresist on the wafer surface that corresponds to the pattern on the mask.

The wafer is then subjected to an etching process, which removes the material not protected by the photoresist. This can be done using either wet etching, which involves immersing the wafer in a chemical solution, or dry etching, which uses a plasma to remove the material.

After etching, the remaining photoresist is stripped away, leaving behind the patterned layer. This process of applying photoresist, exposing, developing, etching, and stripping is repeated for each layer of the integrated circuit.

Dopants are introduced into the silicon wafer using either diffusion or ion implantation. Diffusion involves heating the wafer in a furnace containing a gas with the desired dopant atoms, which then diffuse into the silicon. Ion implantation uses a beam of high-energy ions to implant the dopant atoms directly into the wafer.

Metal layers are deposited on the wafer using techniques like evaporation or sputtering. These layers are then patterned using photolithography and etching to create the interconnections between the different components.

The fabrication of a modern integrated circuit can involve hundreds of individual steps, each of which must be performed with extreme precision. The entire process can take several weeks or even months to complete.

Once the fabrication process is complete, the wafer is tested to identify any defective chips. Each chip is then cut from the wafer, packaged in a protective housing, and tested again before being shipped to customers.

The integrated circuit has had a profound impact on society. It has enabled the development of countless electronic devices that have transformed the way we live, work, and communicate. From personal computers and smartphones to medical devices and the internet, the IC is at the heart of modern technology.

The miniaturization made possible by the integrated circuit has also led to the development of new fields of science and engineering, such as nanotechnology and biotechnology. Scientists are now able to manipulate matter at the atomic and molecular level, creating new materials and devices with unprecedented properties.

The integrated circuit industry is one of the most dynamic and competitive sectors of the global economy. It is characterized by rapid technological change, short product life cycles, and enormous capital investments. Companies in the industry must constantly innovate to stay ahead of the competition, and they must invest billions of dollars in research and development each year.

The development of the integrated circuit was a remarkable achievement, a testament to the ingenuity and perseverance of the scientists and engineers who made it possible. From Kilby's first crude device to the billions-of-transistors chips of today, the IC has come a long way in a relatively short time.

The story of the integrated circuit is far from over. While Moore's Law has slowed in recent years, as the physical limits of transistor scaling are approached, the industry continues to innovate. New materials, such as graphene and carbon nanotubes, are being explored as potential replacements for silicon. New architectures, such as 3D integrated circuits, are being developed to further increase integration density.

Moreover, new types of integrated circuits are being developed for specialized applications. For example, neuromorphic chips, which are designed to mimic the structure and function of the human brain, are being developed for artificial intelligence applications. Quantum computing, which uses the principles of quantum mechanics to perform computations, is another area of active research that could revolutionize computing in the future.

The integrated circuit, the tiny chip of silicon that contains billions of transistors, has transformed our world in countless ways. It has enabled the digital revolution, fueled the growth of the internet, and made possible the development of technologies that were once the stuff of science fiction. It is a story of relentless innovation, driven by the desire to create ever more powerful and sophisticated electronic devices. The journey from the first simple integrated circuits to today's complex systems-on-a-chip is a testament to the power of human ingenuity to reshape the world, one transistor at a time, and to create a future where the only limit to progress is the human imagination itself. The integrated circuit is not just a technical marvel; it is a symbol of the transformative power of science and engineering, a symbol that will undoubtedly continue to evolve and shape our world in the years to come. It is fair to say that the invention of the integrated circuit has been one of the single most important inventions in the history of technology, having driven progress in computing, telecommunications, medical technology, and many other fields. Without the IC, the modern world as we know it would not exist.

CHAPTER TWENTY-TWO: The Power Grid: Distributing Electricity Across Nations

The late 19th and early 20th centuries witnessed a remarkable transformation in the way electricity was generated and used. The invention of practical generators and motors, the development of incandescent lighting, and the triumph of alternating current over direct current had laid the foundation for the widespread adoption of electric power. However, the initial applications of electricity were largely localized. Power plants were typically built to serve individual factories, neighborhoods, or at most, small cities. The idea of a vast, interconnected network of power lines, capable of transmitting electricity over long distances and serving entire regions or nations, was still a distant dream.

This dream would gradually become a reality with the development of the power grid, a complex and interconnected system of power plants, transmission lines, substations, and distribution networks that would revolutionize the way electricity was produced and consumed. The power grid would become one of the most important and complex technological achievements of the 20th century, a vast electrical web that would underpin modern society and enable the unprecedented economic growth and technological advancement of the modern era.

The early electrical systems were mostly isolated and small-scale. Edison's first commercial power plant, Pearl Street Station, which opened in New York City in 1882, served only a small area of lower Manhattan. It used direct current (DC) generators and a network of underground cables to deliver electricity to a few hundred customers within a radius of about a mile.

The limitations of DC for long-distance transmission meant that power plants had to be located close to consumers. This led to the construction of numerous small power plants in urban areas, each serving its own local network. While this approach worked reasonably well for densely populated cities, it was not practical for serving larger areas or for transmitting power from remote hydroelectric plants to urban centers.

The development of alternating current (AC) technology, particularly the invention of the transformer, changed the equation. The ability to step up AC voltage to high levels for efficient transmission and then step it back down for safe use in homes and businesses made it possible to transmit electricity over much longer distances. This opened up the possibility of building larger, more efficient power plants that could serve wider areas.

One of the first major demonstrations of the potential of AC for long-distance transmission came with the construction of the hydroelectric power plant at Niagara Falls in the 1890s. This project, which used Tesla's polyphase AC system, was a landmark achievement. It demonstrated that large amounts of electricity could be generated from a remote location and transmitted over 20 miles to Buffalo, New York, using high-voltage AC transmission lines.

The success of the Niagara Falls project spurred the development of other large-scale hydroelectric plants and the construction of longer transmission lines. It also encouraged the consolidation of smaller, isolated electrical systems into larger, interconnected networks.

The early power grids were often regional in scope, serving a particular city or a group of neighboring towns. These regional grids were typically operated by a single utility company, which owned the power plants, transmission lines, and distribution networks within its service area.

The advantages of interconnected systems soon became apparent. By linking multiple power plants together, utilities could improve the reliability and efficiency of their operations. If one plant experienced a problem or needed to be taken offline for maintenance, other plants in the system could pick up the slack. Interconnection also allowed utilities to share reserve capacity, reducing the need for each plant to have its own backup generators.

As the benefits of interconnection became clear, regional grids began to expand and merge. Utilities started building transmission lines to connect their systems with those of neighboring companies, creating larger and more integrated networks. This process was often driven by economic factors, as larger grids allowed for economies of scale in power generation and transmission.

The growth of the power grid was also facilitated by advances in technology. Higher-voltage transmission lines were developed, allowing for the transmission of electricity over even longer distances with reduced losses. New types of transformers, insulators, and switchgear were developed to handle the higher voltages and currents.

The development of the power grid was not a centrally planned or coordinated process. It evolved organically, driven by the decisions of individual utility companies, technological advancements, and economic forces. However, governments also played a role in shaping the development of the grid, particularly in the areas of regulation and standardization.

In the early days of the electric power industry, there was little regulation or standardization. Each utility company developed its own systems and standards, often with little regard for compatibility with neighboring systems. This led to a patchwork of different voltages, frequencies, and equipment designs, making interconnection difficult and inefficient.

As the grid grew, the need for standardization became increasingly apparent. In the United States, organizations like the American Institute of Electrical Engineers (AIEE) and the National Electric Light Association (NELA) played a key role in developing industry standards for electrical equipment and practices. These standards helped to ensure compatibility between different systems and facilitated the interconnection of regional grids.

Governments also began to regulate the electric power industry, particularly in the areas of safety and reliability. In the United States, state public utility commissions were established to oversee the operations of electric utilities, set rates, and ensure the provision of reliable service. At the federal level, the Federal Power Commission (FPC) was created in 1920 to regulate interstate electricity sales and hydroelectric projects.

The growth of the power grid accelerated in the mid-20th century, driven by increasing demand for electricity, advances in technology, and government policies. The post-World War II economic boom led to a surge in electricity consumption, as households acquired more electrical appliances and industries expanded their use of electric power.

In the United States, the Rural Electrification Administration (REA), established in 1935 as part of President Franklin D. Roosevelt's New Deal, played a crucial role in extending the grid to rural areas. The REA provided low-interest loans to rural electric cooperatives, enabling them to build power lines and provide electricity to farmers and other rural residents who were not served by private utilities.

The development of nuclear power in the 1950s and 1960s also had a significant impact on the grid. Nuclear power plants were much larger than existing fossil fuel plants, and they required high-voltage transmission lines to connect them to the grid. The construction of these large nuclear plants further spurred the development of interconnected systems and the expansion of the high-voltage transmission network.

By the 1960s, the power grid in the United States had evolved into a vast, interconnected network, linking thousands of power plants and millions of consumers across the country. Similar developments were taking place in other industrialized nations, as the benefits of interconnected systems became universally recognized.

The structure of the power grid can be broadly divided into three main components: generation, transmission, and distribution.

Generation is the process of producing electricity from various energy sources, such as fossil fuels (coal, oil, and natural gas), nuclear fuel, hydroelectric power, and renewable sources like wind and solar. Power plants are typically located near sources of fuel or energy, such as coal mines, rivers, or windy areas.

Transmission is the process of transporting electricity over long distances from power plants to substations near populated areas. Transmission lines operate at very high voltages, typically ranging from 69,000 volts (69 kV) to over 765,000 volts (765 kV). High voltages are used to reduce transmission losses, as the power lost in a transmission line is proportional to the square of the current.

Substations are facilities that transform the voltage of electricity from high transmission levels to lower distribution levels. They contain transformers, circuit breakers, and other equipment used to control and protect the flow of electricity.

Distribution is the process of delivering electricity from substations to individual consumers. Distribution lines operate at lower voltages, typically ranging from 4,000 volts (4 kV) to 35,000 volts (35 kV). These lines carry electricity to neighborhoods, where it is stepped down to even lower voltages (e.g., 120/240 volts in the United States) for use in homes and businesses.

The operation of the power grid is a complex and highly coordinated process. Grid operators must constantly monitor the flow of electricity throughout the system, balancing supply and demand in real time. They use sophisticated computer systems and communication networks to track the output of power plants, the status of transmission lines, and the level of electricity consumption across the grid.

One of the key challenges in operating the grid is maintaining the frequency of the alternating current at a constant level. In North

America, the standard frequency is 60 Hz, while in most of the rest of the world it is 50 Hz. If the frequency deviates too far from the standard, it can damage equipment and cause blackouts.

To maintain the frequency, grid operators must ensure that the amount of electricity being generated always matches the amount being consumed. This is a challenging task, as electricity demand can fluctuate significantly throughout the day and across seasons. Grid operators use a variety of techniques to manage these fluctuations, such as adjusting the output of power plants, using energy storage systems, and implementing demand response programs that encourage consumers to reduce their electricity use during peak periods.

The power grid is not a static entity; it is constantly evolving to meet changing needs and incorporate new technologies. One of the major trends in recent years has been the integration of renewable energy sources, such as wind and solar power, into the grid.

Renewable energy sources present both opportunities and challenges for the grid. On the one hand, they offer a clean and sustainable alternative to fossil fuels, helping to reduce greenhouse gas emissions and mitigate climate change. On the other hand, they are intermittent and variable, meaning that their output fluctuates depending on weather conditions.

The intermittency of renewable energy sources poses challenges for grid operators, who must ensure that the grid remains stable and reliable despite the fluctuations in supply. To address these challenges, new technologies and strategies are being developed, such as advanced forecasting systems that can predict wind and solar output, energy storage systems that can store excess electricity for later use, and smart grid technologies that can help to manage the flow of electricity more effectively.

Another major trend is the decentralization of the grid. Traditionally, the grid has been a one-way system, with electricity flowing from large, centralized power plants to consumers. However, with the growth of distributed generation, such as rooftop solar panels and small-scale wind turbines, electricity is increasingly being generated closer to where it is consumed.

This shift towards a more decentralized grid has significant implications for the way the grid is operated and managed. It requires new approaches to grid planning and control, as well as new technologies for integrating and managing distributed energy resources.

The power grid is also becoming more digital and interconnected. The deployment of sensors, communication networks, and advanced control systems is transforming the grid into a "smart grid," a more flexible, responsive, and resilient system that can better integrate renewable energy sources, accommodate electric vehicles, and empower consumers to manage their energy use more effectively.

The development of the power grid has been one of the most significant technological achievements of the modern era. It has enabled the widespread use of electricity, powering homes, businesses, and industries around the world. It has transformed the way we live and work, driving economic growth and improving the quality of life for billions of people.

The story of the power grid is a testament to the power of human ingenuity and the transformative potential of technology. From its humble beginnings as a collection of isolated local systems to its current status as a vast, interconnected network spanning continents, the power grid has come a long way in a relatively short time.

As we look to the future, the power grid will undoubtedly continue to evolve and adapt to new challenges and opportunities. The integration of renewable energy, the decentralization of generation, and the digitalization of the grid are just a few of the trends that are reshaping the electrical landscape.

The power grid of the future will likely be more complex, more dynamic, and more interconnected than ever before. It will need to be more flexible and resilient, capable of accommodating a wide range of energy sources and technologies. It will also need to be more intelligent, using advanced sensors, communication networks, and control systems to optimize its performance and ensure its reliability.

The ongoing evolution of the power grid presents both challenges and opportunities. It requires significant investments in infrastructure, the development of new technologies, and the adoption of new policies and regulations. However, it also offers the potential to create a cleaner, more sustainable, and more equitable energy future. The choices we make today about how we generate, transmit, and use electricity will shape the world we live in for generations to come. The ongoing development and improvement of the power grid is not just a technical challenge; it is a societal imperative, one that will require the combined efforts of engineers, policymakers, business leaders, and citizens around the world.

CHAPTER TWENTY-THREE: Renewable Energy: Harnessing the Power of Nature

The power grid, as we explored in the previous chapter, revolutionized the way electricity was generated, transmitted, and consumed. It enabled the widespread use of electric power, transforming societies and driving economic growth. However, for much of its history, the grid was largely powered by fossil fuels—coal, oil, and natural gas—which, while relatively abundant and energy-dense, are finite resources whose extraction and combustion have significant environmental consequences.

The burning of fossil fuels releases greenhouse gases, such as carbon dioxide, into the atmosphere. These gases trap heat, contributing to global warming and climate change. The extraction and transportation of fossil fuels can also cause environmental damage, such as habitat destruction, oil spills, and water pollution.

As the environmental impacts of fossil fuels became increasingly apparent in the latter half of the 20th century, a growing movement emerged to develop and deploy alternative sources of energy that were cleaner, more sustainable, and less damaging to the planet. This movement has led to the rise of renewable energy, a diverse array of technologies that harness the power of natural processes to generate electricity.

Renewable energy sources, such as solar, wind, hydro, geothermal, and biomass, are fundamentally different from fossil fuels. They are naturally replenished over relatively short periods, and their use does not deplete finite resources. Moreover, they produce little or no greenhouse gas emissions during operation, making them a key part of the global effort to combat climate change.

The development of renewable energy technologies has been driven by a combination of factors, including environmental concerns, technological advancements, government policies, and economic forces. While renewable energy still accounts for a relatively small share of global electricity generation compared to fossil fuels, its growth has been rapid in recent years, and it is poised to play an increasingly important role in the energy mix of the future.

One of the most prominent forms of renewable energy is solar power. Solar power harnesses the energy of the sun, the most abundant energy source on Earth, and converts it into electricity. There are two main types of solar power technologies: photovoltaic (PV) and concentrated solar power (CSP).

Photovoltaic (PV) technology uses solar cells to directly convert sunlight into electricity. Solar cells are made of semiconductor materials, typically silicon, that exhibit the photovoltaic effect. When sunlight strikes a solar cell, photons (particles of light) are absorbed by the semiconductor material. The energy from these photons excites electrons in the material, causing them to flow, creating an electric current.

The first practical solar cell was developed at Bell Labs in 1954, the same laboratory where the transistor was invented. This early solar cell had an efficiency of about 4%, meaning that it could convert 4% of the incident sunlight into electricity. While this was a significant breakthrough, it was not yet efficient enough for widespread use.

Over the following decades, researchers around the world worked to improve the efficiency and reduce the cost of solar cells. They experimented with different semiconductor materials, such as gallium arsenide and cadmium telluride, and developed new cell designs, such as multi-junction cells, which use multiple layers of different materials to capture a broader spectrum of sunlight.

These efforts have led to significant improvements in solar cell performance. Today, the most efficient commercially available solar panels have efficiencies of over 20%, and some experimental cells have achieved efficiencies of over 40% in laboratory settings.

The cost of solar PV has also declined dramatically. In the 1970s, solar panels were extremely expensive, costing over $100 per watt of capacity. However, driven by technological advancements, economies of scale, and increased competition, the cost has plummeted. Today, the cost of solar panels is less than $0.50 per watt, and the cost of electricity from large-scale solar installations is competitive with that of fossil fuels in many parts of the world.

The rapid decline in the cost of solar PV has fueled a boom in solar installations. Solar panels are now being deployed on rooftops, in large-scale solar farms, and even in space. They are being used to power homes, businesses, and entire communities. In many developing countries, solar power is providing electricity to areas that are not connected to the grid, leapfrogging the need for expensive and time-consuming grid expansion.

Concentrated solar power (CSP) is another type of solar technology that uses mirrors to focus sunlight onto a receiver, which then heats a fluid. This hot fluid is used to generate steam, which drives a turbine to produce electricity. CSP plants are typically large-scale installations that are best suited for areas with high levels of direct sunlight.

CSP technology has not seen the same dramatic cost reductions as solar PV, but it has the advantage of being able to store heat, allowing for electricity generation even when the sun is not shining. Some CSP plants use molten salt as a heat storage medium, which can retain heat for several hours, enabling the plant to continue producing electricity after sunset.

Another major source of renewable energy is wind power. Wind power harnesses the kinetic energy of moving air and converts it into electricity using wind turbines. Wind turbines typically consist of a tower, a nacelle that houses the generator and other equipment, and large blades that capture the wind's energy.

The basic principle of wind power is simple: as wind flows over the blades of a turbine, it creates lift, similar to the lift that allows an airplane to fly. This lift causes the blades to rotate, and the rotation drives a generator that produces electricity.

The first electricity-generating wind turbine was built in 1887 by Scottish academic James Blyth, but it would take nearly a century for wind power to become a significant source of electricity. In the late 20th and early 21st centuries, driven by concerns about climate change and advances in turbine technology, wind power experienced a period of rapid growth.

Modern wind turbines are marvels of engineering. They can be over 100 meters tall, with blades that span over 60 meters. They are equipped with sophisticated control systems that adjust the pitch of the blades and the orientation of the nacelle to maximize energy capture and protect the turbine from damage in high winds.

Wind turbines can be installed on land (onshore wind) or in bodies of water (offshore wind). Onshore wind is currently the more common and less expensive option, but offshore wind has the potential to generate more electricity due to the stronger and more consistent winds found over water.

The cost of wind power has declined significantly in recent decades, driven by improvements in turbine technology, economies of scale, and increased competition. Today, wind power is one of the most cost-competitive sources of new electricity generation, and it is being deployed on a large scale around the world.

Wind power, like solar power, is an intermittent energy source, meaning that its output varies depending on weather conditions. This intermittency poses challenges for grid operators, who must ensure a stable and reliable supply of electricity at all times. However, as wind power's share of the energy mix grows, new strategies and technologies are being developed to manage its variability, such as advanced weather forecasting, energy storage, and smart grid technologies.

Hydropower is another important source of renewable energy. Hydropower harnesses the potential energy of water stored in dams or the kinetic energy of flowing water in rivers and converts it into electricity. Hydroelectric plants use turbines and generators, similar to those used in wind turbines, to convert the energy of moving water into electricity.

Hydropower is the oldest and most established form of renewable energy. The first hydroelectric plant was built in 1882 in Appleton, Wisconsin, and hydropower has been a significant source of electricity ever since. Today, hydropower accounts for the largest share of global renewable electricity generation.

There are two main types of hydropower plants: storage hydropower and run-of-river hydropower. Storage hydropower plants use a dam to create a reservoir, storing water and releasing it through turbines to generate electricity. Run-of-river plants, on the other hand, divert a portion of a river's flow through turbines without creating a large reservoir.

Hydropower has several advantages as a renewable energy source. It is a mature and well-understood technology, with high efficiency and low operating costs. It is also a dispatchable source of energy, meaning that its output can be adjusted to meet changes in demand. Large hydropower plants with reservoirs can also provide flood control and water storage benefits.

However, hydropower also has environmental and social impacts. Large dams can alter river ecosystems, affecting fish migration and water quality. They can also displace communities and inundate valuable land. The construction of large dams can also be controversial, with concerns about their impact on local communities and the environment.

Run-of-river hydropower projects generally have a smaller environmental footprint than large storage projects, but they can still have impacts on river ecosystems. Careful planning and design are needed to minimize the environmental and social impacts of hydropower projects.

Geothermal energy is another form of renewable energy that harnesses the heat from the Earth's interior to generate electricity. Geothermal power plants use wells to access hot water or steam reservoirs beneath the Earth's surface. The hot water or steam is then used to drive turbines and generate electricity.

Geothermal energy is a reliable and continuous source of power, as the Earth's internal heat is constantly being replenished. It is also a relatively clean energy source, with low greenhouse gas emissions compared to fossil fuels.

However, geothermal resources are not evenly distributed around the world. They are most abundant in regions with high levels of tectonic activity, such as Iceland, New Zealand, and parts of the western United States. The development of geothermal power plants also requires careful management of the geothermal reservoir to ensure its long-term sustainability.

Biomass is another renewable energy source that involves burning organic matter, such as wood, agricultural residues, and municipal solid waste, to generate heat or electricity. Biomass can be used in power plants similar to those that burn fossil fuels, or it can be converted into biofuels, such as ethanol and biodiesel, for use in transportation.

Biomass is often considered a carbon-neutral energy source, as the carbon released during combustion is offset by the carbon absorbed by the plants during their growth. However, the sustainability of biomass depends on how it is sourced and managed. If biomass is harvested from unsustainable sources, such as old-growth forests, or if its production competes with food crops, it can have negative environmental and social impacts.

The growth of renewable energy has been driven by a combination of factors. Environmental concerns, particularly the need to reduce greenhouse gas emissions and mitigate climate change, have been a major driver. Technological advancements have also played a crucial role, making renewable energy more efficient and cost-competitive.

Government policies have also been instrumental in promoting the development and deployment of renewable energy. Many countries have implemented policies such as feed-in tariffs, which guarantee a fixed price for renewable electricity, and renewable portfolio standards, which require utilities to generate a certain percentage of their electricity from renewable sources. Tax incentives, grants, and other financial support mechanisms have also been used to encourage investment in renewable energy projects.

The economics of renewable energy have also shifted dramatically in recent years. As the cost of solar and wind power has plummeted, they have become increasingly competitive with fossil fuels, even without subsidies. In many parts of the world, new solar and wind projects are now the cheapest source of electricity generation, making them an attractive option for utilities and investors.

The growth of renewable energy is transforming the electricity sector. It is leading to a more diverse and decentralized energy mix, with a greater reliance on distributed generation and smart grid technologies. It is also creating new opportunities for economic development and job creation, as the renewable energy industry grows and matures.

However, the transition to a renewable energy future is not without its challenges. The intermittency of solar and wind power requires new approaches to grid management and operation. The integration of large amounts of variable renewable energy requires upgrades to transmission infrastructure, the deployment of energy storage technologies, and the development of more flexible and responsive grid systems.

The rapid growth of renewable energy also raises questions about the future of existing fossil fuel infrastructure. As renewable energy becomes more cost-competitive, there is a risk that fossil fuel assets, such as coal-fired power plants, could become "stranded," meaning that they are no longer economically viable and must be retired early. This could have significant financial implications for utilities and investors, and it underscores the need for careful planning and management of the energy transition.

Despite these challenges, the shift towards renewable energy is gaining momentum. The costs of solar and wind power continue to fall, and new technologies, such as advanced energy storage and smart grid systems, are being developed and deployed at an increasing pace.

Many countries have set ambitious targets for renewable energy deployment and greenhouse gas emissions reductions. The Paris Agreement, adopted in 2015, aims to limit global warming to well below 2 degrees Celsius above pre-industrial levels, and countries around the world have committed to reducing their emissions as part of this agreement.

The transition to a renewable energy future is not just an environmental imperative; it is also an economic opportunity. The renewable energy industry is creating jobs and driving innovation. It is also enhancing energy security by reducing reliance on imported fossil fuels and diversifying the energy mix.

The rise of renewable energy is a transformative moment in the history of electricity. It represents a fundamental shift in the way we generate and use power, a shift that is driven by the need to address climate change, improve air quality, and create a more sustainable and resilient energy system.

The journey towards a renewable energy future is a complex and multifaceted one. It requires technological innovation, policy support, and changes in consumer behavior. It also requires a long-term vision and a commitment to investing in the infrastructure and technologies that will be needed to support a clean energy economy.

The story of renewable energy is still being written. The choices we make today about how we generate and use electricity will determine the kind of world we leave to future generations. The rise of renewable energy offers a pathway to a cleaner, more sustainable, and more prosperous future, a future where the power of nature is harnessed for the benefit of all. It is a future where the sun, the wind, the rivers, and the Earth itself provide the energy we need to power our homes, our businesses, and our lives, without compromising the health of our planet or the well-being of future generations. It is up to all of us to make that future a reality.

CHAPTER TWENTY-FOUR: Superconductivity: The Quest for Zero Resistance

The discovery of superconductivity in the early 20th century marked a significant milestone in our understanding of electricity and materials science. This phenomenon, characterized by the complete disappearance of electrical resistance in certain materials at extremely low temperatures, opened up a new frontier in physics and engineering, promising revolutionary applications in power transmission, magnetic levitation, and advanced electronics.

The story of superconductivity begins in 1908 with the work of Dutch physicist Heike Kamerlingh Onnes. Onnes had established a cryogenics laboratory at the University of Leiden, where he was pursuing research into the behavior of materials at very low temperatures. In 1908, he became the first person to liquefy helium, achieving a temperature of just 4.2 Kelvin (-268.95°C or -452.11°F), the lowest temperature attained on Earth at that time.

With this breakthrough, Onnes was able to study the electrical properties of materials at temperatures approaching absolute zero. One of the questions he sought to answer was how the electrical resistance of metals would change as they approached this ultimate low temperature. There were several competing theories at the time. Some scientists believed that the resistance would gradually decrease to zero, while others thought it might increase indefinitely, effectively turning the metal into an insulator.

In 1911, Onnes and his team were investigating the resistance of mercury at low temperatures. Mercury was chosen because it could be purified to a very high degree and, being liquid at room temperature, could be easily shaped into wires. As they lowered the temperature, they observed the expected decrease in resistance. However, when the temperature reached 4.2 K, something unexpected happened: the resistance suddenly disappeared entirely.

Onnes initially thought there might be a problem with his equipment, but repeated experiments confirmed the result. He had discovered a new state of matter, which he called the "superconducting state." This discovery earned Onnes the Nobel Prize in Physics in 1913 "for his investigations on the properties of matter at low temperatures which led, inter alia, to the production of liquid helium."

In the years following Onnes' discovery, researchers found that many other metals and alloys also became superconducting at low temperatures. Each material had its own critical temperature (Tc) at which the transition to the superconducting state occurred. However, all of these early superconductors required extremely low temperatures, typically below 20 K, which limited their practical applications.

The next major breakthrough in superconductivity came in 1933 when German physicists Walther Meissner and Robert Ochsenfeld discovered a unique property of superconductors. They found that when a superconductor is placed in a weak external magnetic field, it expels the field from its interior as it transitions into the superconducting state. This phenomenon, known as the Meissner effect, became a defining characteristic of superconductivity.

The Meissner effect distinguishes superconductors from perfect conductors. While a perfect conductor would allow a magnetic field to pass through it unchanged, a superconductor actively expels the field, regardless of whether the field was applied before or after the material became superconducting. This property is the basis for magnetic levitation using superconductors, where a magnet can be made to float above a superconductor due to the repulsion of the magnetic field.

Despite these discoveries, a theoretical understanding of superconductivity remained elusive for decades. It wasn't until 1957 that a comprehensive theory was developed by American physicists John Bardeen, Leon Cooper, and John Robert Schrieffer. Their theory, known as BCS theory (after their initials), explained superconductivity as a quantum mechanical effect.

According to BCS theory, electrons in a superconductor form pairs, called Cooper pairs, through interactions with vibrations in the crystal lattice of the material. These Cooper pairs can move through the material without encountering resistance, as they are able to pass through the vibrating lattice without scattering. The theory successfully explained many of the observed properties of superconductors and predicted new phenomena, earning Bardeen, Cooper, and Schrieffer the Nobel Prize in Physics in 1972.

While BCS theory provided a fundamental understanding of superconductivity, it also suggested limitations. The theory implied that superconductivity was unlikely to occur at temperatures much above 30 K. This prediction seemed to be borne out by experiments, as the highest Tc achieved by the mid-1980s was only 23.2 K for a niobium-germanium compound.

However, in 1986, a dramatic breakthrough occurred that would challenge the limits suggested by BCS theory. IBM researchers Georg Bednorz and K. Alex Müller discovered superconductivity in a lanthanum-barium-copper oxide (LBCO) ceramic at a temperature of 35 K. This was significantly higher than any previously known superconductor, and it opened up a new class of materials called high-temperature superconductors.

The discovery by Bednorz and Müller sparked a flurry of research activity around the world. Within months, researchers had found related copper oxide compounds with even higher critical temperatures. In 1987, a yttrium-barium-copper oxide (YBCO) compound was found to become superconducting at 93 K, a temperature above the boiling point of liquid nitrogen (77 K). This was a significant milestone, as liquid nitrogen is much cheaper and easier to work with than liquid helium.

The advent of high-temperature superconductors brought the possibility of practical applications much closer to reality. While 93 K is still very cold by everyday standards, it's much more achievable than the near-absolute zero temperatures required by earlier superconductors. This opened up new possibilities for using superconductors in power transmission, magnetic resonance imaging (MRI) machines, and other applications.

However, high-temperature superconductors also presented new challenges. Unlike conventional superconductors, which are typically simple metals or alloys, high-temperature superconductors are complex ceramic materials with layered crystal structures. They are often brittle and difficult to form into wires or other useful shapes. Moreover, their behavior is not fully explained by BCS theory, and developing a comprehensive theory of high-temperature superconductivity remains one of the great challenges in condensed matter physics.

Despite these challenges, research into high-temperature superconductivity has continued apace. In the years since the discovery of YBCO, researchers have found numerous other high-temperature superconductors, including mercury-based compounds with Tc as high as 138 K under normal pressure, and up to 166 K under high pressure.

In 2015, another breakthrough occurred with the discovery of superconductivity in hydrogen sulfide under extremely high pressure. When compressed to about 1.5 million atmospheres, hydrogen sulfide was found to become superconducting at temperatures as high as 203 K (-70°C or -94°F). While the extreme pressures required make this particular material impractical for most applications, it demonstrated that superconductivity at even higher temperatures might be possible.

The quest for room-temperature superconductivity, long considered a "holy grail" of materials science, has gained new momentum in recent years. In 2018, researchers reported superconductivity at -23°C (250 K) in lanthanum hydride under high pressure. In 2019, a compound of lanthanum and hydrogen was reported to superconduct at even higher temperatures, up to -13°C (260 K), albeit still under extreme pressure.

These discoveries have reignited interest in hydride compounds as potential high-temperature superconductors. Some researchers believe that metallic hydrogen, which is predicted to exist at extremely high pressures, could be a room-temperature superconductor. However, creating and studying materials under such extreme conditions presents enormous technical challenges.

While the search for room-temperature superconductors continues, existing superconductors have already found important applications in various fields. One of the most significant is in the field of medical imaging. Magnetic Resonance Imaging (MRI) machines use powerful superconducting magnets to generate the strong, uniform magnetic fields needed for high-resolution imaging. The use of superconductors allows these machines to create much stronger magnetic fields than would be possible with conventional electromagnets.

Superconductors are also used in particle accelerators, such as the Large Hadron Collider at CERN. These machines use superconducting magnets to steer and focus beams of high-energy particles. The strong magnetic fields produced by superconducting magnets allow particles to be accelerated to much higher energies than would be possible with conventional magnets.

In the field of energy technology, superconductors have potential applications in power transmission and energy storage. Superconducting power lines could significantly reduce energy losses in long-distance power transmission. While such systems are not yet widely deployed, several demonstration projects have shown their potential. For example, a superconducting cable system has been operating in the power grid of Essen, Germany, since 2014.

Superconductors are also being explored for use in energy storage devices called Superconducting Magnetic Energy Storage (SMES) systems. These systems store energy in the magnetic field of a superconducting coil. While current SMES systems are mainly used for short-term energy storage to improve power quality, larger systems could potentially be used for grid-scale energy storage in the future.

Another promising application of superconductivity is in the development of quantum computers. Many designs for quantum computers rely on superconducting circuits to create and manipulate quantum bits, or qubits. The unique properties of superconductors, particularly their ability to maintain quantum coherence over relatively long time scales, make them attractive for this application.

Superconductivity also plays a crucial role in some of the most sensitive detectors ever created. Superconducting quantum interference devices (SQUIDs) can measure incredibly weak magnetic fields, with applications ranging from brain imaging to the search for dark matter in cosmology.

Despite these successes, many challenges remain in the field of superconductivity. The need for cryogenic cooling, even for high-temperature superconductors, remains a significant barrier to widespread adoption in many applications. The brittleness of many high-temperature superconductors makes them difficult to form into wires or other useful shapes. And the fundamental mechanisms behind high-temperature superconductivity are still not fully understood, hampering efforts to design new superconducting materials rationally.

Researchers are approaching these challenges from multiple angles. Materials scientists are exploring new compounds and structures that might exhibit superconductivity at higher temperatures or under less extreme conditions. Engineers are developing new techniques for fabricating and shaping superconducting materials. And physicists are working to develop more comprehensive theories of superconductivity that can explain the behavior of both conventional and high-temperature superconductors.

One promising area of research is in two-dimensional superconductors. In recent years, researchers have discovered superconductivity in atomically thin layers of certain materials, including graphene (when doped with calcium) and certain transition metal dichalcogenides. These 2D superconductors exhibit unique properties and may offer new insights into the mechanisms of superconductivity.

Another active area of research is in topological superconductors. These materials combine superconductivity with topological properties, which relate to the geometric structure of the material's electronic states. Topological superconductors are of particular interest because they may host exotic particles called Majorana fermions, which could potentially be used to create quantum computers that are inherently resistant to errors.

The field of superconductivity continues to surprise and challenge researchers more than a century after its initial discovery. Each new breakthrough brings us closer to understanding this remarkable phenomenon and harnessing its potential. While room-temperature superconductivity remains an elusive goal, the progress made in recent years gives reason for optimism.

The quest for superconductivity at ever-higher temperatures is not just a scientific curiosity. It has the potential to revolutionize our energy infrastructure, our computing capabilities, and our ability to probe the fundamental nature of matter. As we continue to push the boundaries of what's possible with superconductors, we may find ourselves on the brink of a new technological revolution, one that could be as transformative as the original electrical revolution of the 19th and 20th centuries.

The story of superconductivity is a testament to the power of basic scientific research. What began as a curiosity in a low-temperature laboratory has grown into a field with profound implications for technology and our understanding of the physical world. It reminds us that in science, as in many areas of human endeavor, the most revolutionary discoveries often come from exploring the extremes and questioning our fundamental assumptions about how the world works.

As we look to the future, the field of superconductivity continues to hold great promise. Whether through the discovery of new high-temperature superconductors, the development of novel applications for existing materials, or breakthroughs in our theoretical understanding, superconductivity is likely to remain at the forefront of physics and materials science for years to come. The dream of room-temperature superconductivity, and the technological marvels it could enable, continues to inspire researchers around the world, driving us ever closer to a future where the flow of electricity encounters no resistance at all.

CHAPTER TWENTY-FIVE: The Future of Electricity: Smart Grids, Electric Vehicles, and Beyond

As we stand on the brink of a new era in electrical technology, the future of electricity promises to be as transformative as its past. The electrical grid, which has remained largely unchanged for over a century, is poised for a dramatic overhaul. This evolution is driven by the need to integrate renewable energy sources, improve efficiency, and meet the growing demand for electricity in an increasingly electrified world. At the forefront of this transformation are smart grids, electric vehicles, and emerging technologies that promise to revolutionize the way we generate, distribute, and use electricity.

The concept of the smart grid represents a fundamental shift in how we think about electricity distribution. Unlike the traditional one-way power flow from centralized power plants to consumers, a smart grid is a two-way system that allows for real-time communication between utilities and consumers. This bidirectional flow of information and electricity enables more efficient management of power resources, integration of renewable energy sources, and empowerment of consumers to make informed decisions about their energy use.

At the heart of the smart grid are advanced metering infrastructure (AMI) systems, often referred to as smart meters. These devices provide real-time data on electricity consumption, allowing utilities to better manage supply and demand. For consumers, smart meters offer detailed insights into their energy usage patterns, enabling them to make more informed decisions about when and how they use electricity. This information can help consumers reduce their energy bills and contribute to overall grid efficiency.

One of the key advantages of smart grids is their ability to integrate distributed energy resources (DERs) more effectively.

DERs include rooftop solar panels, small wind turbines, energy storage systems, and even electric vehicles. As these resources become more prevalent, the grid needs to be able to accommodate bidirectional power flows and manage the intermittent nature of renewable energy sources.

Smart grids use sophisticated control systems and artificial intelligence to balance supply and demand in real-time. When the sun is shining and solar panels are producing excess electricity, the grid can distribute this power to where it's needed most. Conversely, when renewable sources are not producing enough power, the grid can quickly ramp up other sources or draw from energy storage systems to meet demand.

Energy storage is a crucial component of future smart grids. As renewable energy sources like solar and wind become more prevalent, the need for efficient and cost-effective energy storage solutions becomes more pressing. While lithium-ion batteries have seen significant improvements in recent years, researchers are exploring a variety of other storage technologies, including flow batteries, compressed air energy storage, and even novel approaches like gravity-based storage systems.

Another key aspect of smart grids is their ability to implement demand response programs. These programs incentivize consumers to reduce their electricity usage during peak demand periods, helping to balance the load on the grid and reduce the need for expensive peaker plants. With smart appliances and home energy management systems, consumers can automatically adjust their electricity usage based on real-time pricing signals from the utility.

The transition to smart grids also involves hardening the infrastructure against physical and cyber threats. As the grid becomes more reliant on digital technologies and interconnected systems, it becomes more vulnerable to cyberattacks. Utilities and government agencies are investing heavily in cybersecurity measures to protect this critical infrastructure.

Alongside the development of smart grids, the rise of electric vehicles (EVs) is set to have a profound impact on the future of electricity. As countries around the world set ambitious targets to phase out internal combustion engines, the demand for EVs is expected to surge in the coming decades. This shift will not only reduce greenhouse gas emissions from the transportation sector but also create new challenges and opportunities for the electrical grid.

The widespread adoption of EVs will significantly increase electricity demand, particularly during charging periods. However, with smart charging systems, EVs can become a valuable asset to the grid rather than a burden. Vehicle-to-grid (V2G) technology allows EVs to not only draw power from the grid but also feed power back when needed. This turns EVs into mobile energy storage units, helping to balance the grid and potentially providing backup power during outages.

Smart charging systems can schedule EV charging during off-peak hours when electricity is cheaper and more abundant. This helps to flatten the demand curve, reducing strain on the grid and potentially lowering costs for consumers. Some utilities are already experimenting with time-of-use pricing and other incentives to encourage off-peak charging.

The development of fast-charging infrastructure is another crucial aspect of the EV revolution. While most EV charging currently happens at home or work, the availability of fast-charging stations along highways and in urban areas is essential for long-distance travel and to alleviate "range anxiety." Companies and governments around the world are investing heavily in expanding charging networks, with some ambitious projects aiming to install chargers every few miles along major highways.

As battery technology continues to improve, we can expect to see EVs with longer ranges, faster charging times, and lower costs. Solid-state batteries, which promise higher energy density and improved safety compared to current lithium-ion batteries, are on the horizon. If successfully commercialized, solid-state batteries could dramatically accelerate the adoption of EVs.

The electrification of transportation extends beyond personal vehicles. Electric buses are already becoming common in many cities, offering lower operating costs and reduced emissions compared to diesel buses. The electrification of trucking is also gaining momentum, with several companies developing electric semi-trucks for long-haul transport. Even air travel may eventually be electrified, with several startups working on electric aircraft for short-haul flights.

Looking further into the future, we can anticipate several emerging technologies that could reshape the electrical landscape. Wireless power transmission, once the dream of Nikola Tesla, is slowly becoming a reality. While currently limited to short-range applications like charging pads for smartphones, researchers are exploring ways to transmit power over longer distances. If successful, this technology could eliminate the need for power cords and potentially even allow for the wireless charging of electric vehicles while driving.

Nuclear fusion, often described as the "holy grail" of energy production, continues to tantalize scientists with its promise of abundant, clean energy. While fusion power has remained elusive for decades, recent advancements in superconducting magnets and plasma confinement techniques have brought us closer than ever to achieving net energy gain from fusion reactions. If successfully developed, fusion power could provide a virtually limitless source of electricity with minimal environmental impact.

The concept of space-based solar power is also gaining renewed interest. This involves collecting solar energy in orbit, where sunlight is constant and unaffected by weather or day-night cycles, and beaming it back to Earth using microwave or laser technology. While the technical and economic challenges are formidable, space-based solar power could potentially provide a continuous, renewable source of electricity on a global scale.

Quantum technologies may also play a role in the future of electricity. Quantum sensors could enable ultra-precise measurements of electrical fields and currents, improving the efficiency and reliability of power systems. Quantum computing, while still in its infancy, could potentially revolutionize grid management and optimization, solving complex problems that are intractable for classical computers.

As we move towards a more electrified future, the way we think about and interact with electricity is likely to change. The lines between energy consumers and producers will continue to blur, with more individuals and businesses generating their own electricity through solar panels, small wind turbines, or other distributed energy resources. Energy communities and microgrids may become more common, allowing neighborhoods or industrial parks to generate, store, and manage their own electricity, potentially operating independently from the main grid when needed.

The increasing electrification of various sectors of the economy, from transportation to heating and industrial processes, will require significant upgrades to our electrical infrastructure. This includes not only the transmission and distribution networks but also the generation capacity. While renewable energy sources will play an increasingly important role, we may also see a resurgence of interest in next-generation nuclear power, including small modular reactors and advanced fission designs, as a low-carbon baseload power source.

The future of electricity will also be shaped by advancements in materials science. New materials with improved conductivity, such as high-temperature superconductors, could dramatically reduce transmission losses and enable more efficient electrical devices. Novel semiconductor materials and designs could lead to more efficient solar cells, power electronics, and lighting systems.

As our electrical systems become more complex and interconnected, the role of artificial intelligence and machine learning in managing these systems will become increasingly important. AI algorithms could optimize power flow across the grid, predict and prevent equipment failures, and manage the integration of diverse energy sources and storage systems.

The electrification of the developing world presents both challenges and opportunities. As billions of people gain access to electricity for the first time, there's an opportunity to leapfrog older technologies and build modern, efficient electrical systems from the ground up. Distributed renewable energy systems and microgrids could bring electricity to remote areas without the need for extensive transmission infrastructure.

The future of electricity is inextricably linked to the broader challenge of climate change. As we strive to decarbonize our energy systems, electricity will play a central role in replacing fossil fuels across various sectors of the economy. This transition will require not only technological innovations but also policy changes, market reforms, and shifts in consumer behavior.

As we look to the future, it's clear that electricity will continue to be a driving force in technological progress and social development. From the smart grids that will power our cities to the electric vehicles that will transform our transportation systems, from the renewable energy technologies that will help us combat climate change to the emerging technologies that promise to revolutionize power generation and transmission, the future of electricity is bright with possibility. As we continue to innovate and push the boundaries of what's possible, we can look forward to a future where clean, affordable, and reliable electricity is available to all, powering a more sustainable and prosperous world for generations to come

Printed in Great Britain
by Amazon